Statistics of Financial Markets

Statistics of Financial Markets

Editor

Jacek Fabian

scitus
academics

Statistics of Financial Markets
Edited by **Jacek Fabian**

ISBN: 978-1-68117-647-5
Library of Congress Control Number: 2016946510

Notice

Printed in United States of America on Acid Free Paper

Preface

Statistics is used every day in business. It can be used for quality assurance, financial analysis, production and operations, and many other business areas. Until about the 1970s, financial mathematics has been rather modest compared with other mathematical disciplines. This changed rapidly after the path-breaking works of F. Black, M. Scholes, and R. Merton on derivative pricing, for which they received the Nobel prize of economics in 1997. Since 1973, the publication year of the famous Black and Scholes article, the importance of derivative instruments in financial markets has not ceased to grow. Higher risks associated with, for example, flexible instead of fixed exchange rates after the fall of the Bretton Woods system required a risk management and the use of hedging instruments for internationally active companies. More recently, globalization and the increasingly complex dependence of financial markets are reasons for using sophisticated mathematical and statistical methods and models to evaluate risks. As a result, there is an increasing demand for experts in financial engineering, who control risks internally, search for profitable investment opportunities and guarantee the obligations of legislation. In the future, such risk management is likely to become obligatory for other, deregulated markets such as telecommunication and energy markets. Being aware of the increasing price, volume, and credit risks in these markets, large companies usually have already created new departments dealing with asset and liability management as well as risk management.

Preface

Statistics of Financial Markets offers a vivid yet concise introduction to the growing field of statistical applications in finance. The focus is both on fundamentals of mathematical finance and financial time series analysis and on applications to given problems of financial markets, making the book the ideal basis for lectures, researchers, and practitioners.

TABLE OF CONTENTS

1

CHAPTER

Production in General Equilibrium with Incomplete Financial Markets

Pascal Stiefenhofer

Department of Mathematics, University of Sussex, Brighton, UK

ABSTRACT

This paper considers a general equilibrium model with incomplete financial markets where production sets depend on the financial decisions of the firms. In the short run, firms make financial choices in order to build up production capacity. Given production capacity firms make profit maximizing production decisions in period two. We provide the conditions of existence of equilibria.

Keywords: General Equilibrium, Incomplete Financial Markets, Production, Existence of Equilibria, Transversality

1. INTRODUCTION

Classical general equilibrium literature on production with incomplete markets has focused on variations of the Arrow's seminal two-period model with exogenous financial assets [1] [2] . In this framework, the firm's real sequential optimization structure is independent of its financial activities.

Firms choose quantities of inputs of production in period one such that associated output choices in period two are optimal. This concept of the firm corresponds to the private ownership model of the firm introduced by Debreu [3] , where the single argument of the firm's two period sequential optimization function is the real activity vector. In these recent models, influenced by Drèze [4] and Grossmann & Hart [5] , optimality of the choice of a net real activity vector over two periods refers to the average utility of the group of owners of the firm, the stock holders. It is in that sense that the literature has assigned utilities to firms and that the firms' objective is to maximize some average utility of the share holders. The two concepts applied in most models, slightly differ in the choice of average utility utilized (average utility of initial/final share holders). For a sample of the huge literature applying these concepts see [4] - [8] .

This paper introduces a model of the firm, where its financial and real activities are independent of any average utility of the stock holders. It postulates that firms maximize long run profits and make financial and real decisions sequentially over two periods. The assumption of long run profit maximization is justified by the sequential optimization structure of the firm. Firms issue stocks in period one in order to acquire the cash needed to install production capacity. The optimal quantity of stocks issued by each firm is endogenously determined by the model. Once capacity is installed, after uncertain state of nature has occurred at the beginning of period two, firms produce real goods subject to capacity and technological constraints. The ownership structure introduced in this model eliminates the strategic choice problem of the firm present in the literature. Here, stock holders do not decide about the optimal input vector of the firm in period one. They invest in firms by purchasing stocks in order to transfer wealth across time and between uncertain states of nature. The total quantity of stocks demanded is equal to total quantity of stocks supplied by firms in the same period. The value of total stocks issued by a firm bounds the value of inputs a firm can purchase in period two. Real activities of the firm take place after uncertainty in period two has resolved. These production activities correspond to finding the optimal net activity vector at given prices and revealed state of the world such that profits are maximized at given production capacity.

The sine qua non of the model is then to show that equilibrium exists. It is shown that, for an endogenized price and technology dependent real asset structure, which is transverse to the reduced rank manifolds, equili- brium exists generically in the endowments by the application of Thom's parametric transversality theorem. Finally, the non-smooth convex production set case is considered, where the piecewise linear production manifolds are regularized by convolution. Existence then follows from the smooth case. Bottazzi [9] demonstrated generic existence of equilibrium for an exchange economy for price dependent smooth assets. Equilibria exist for more general asset structures.

The model is introduced in Section 2. Section 3 shows generic existence for convex smooth production manifolds.

2. THE MODEL

We consider a two period $t \in \{0,1\}$ model with uncertainty in period 1 represented as states of nature. An element in the set of mutually exclusive and exhaustive uncertain events is denoted $s \in S = \{1,\cdots,S\}$, where by Convention $s=0$ represents the certain event in period 0, and S denotes the set of all mutually exclusive uncertain events. This set denotes the overall description of uncertainty in the model, which is characterized by

idiosyncratic and aggregate risk. The general uncertainty space is described by the Cartesian product $S = \mathcal{S} \times \tilde{\mathcal{S}}$. For every production set Y_j, there exists a set of states of nature $S_j = \{1,\cdots,\mathbb{S}\}$, where $\mathbb{S} \geq 2$, for all S_j. Denote $\mathcal{S} = \{S_1,\cdots,S_j,\cdots,S_n\}$, where $S \subseteq \mathcal{S}$, the set of technological uncertain events. At aggregate level there are $\tilde{\mathcal{S}} = \{1,\cdots,\tilde{\mathcal{S}}\}$ states of nature. We count in total $(S+1)$ states of nature.

The economic agents are the $j \in \{1,\cdots,n\}$ producers and $i \in \{1,\cdots,m\}$ consumers which are characterized by sets of assumptions F and C bellow. There are $k \in \{1,\cdots,l\}$ physical commodities and $j \in \{1,\cdots,n\}$ financial assets, referred to as stocks. Physical goods are traded on each of the $(S+1)$ spot markets. Firms

issue stocks which are traded at $s=0$, yielding a payoff in the next period at uncertain state $s \in \{1,\cdots,S\}$. The quantity vector of stocks issued by firm j is denoted $z_j \in \mathbb{R}_-$. Other assets such as bonds or options can be introduced without any further difficulties. There are total $l(S+1)$ goods. The consumption bundle of agent i is denoted $x_i = (x_i(0), x_i(s), \cdots, x_i(S)) \in \mathbb{R}_{++}^{l(S+1)}$, with

$x_i(s)=(x_i^1(s),\cdots,x_i^l(s))\in\mathbb{R}_{++}^l$, and $\sum_{i=1}^m x_i = x$. The consumption space for each i is $X_i=\mathbb{R}_{++}^{l(S+1)}$, the strictly positive orthant. The associated price system is a collection of vectors represented by $p=(p(0),p(s),\cdots,p(S))\in\mathbb{R}_{++}^{l(S+1)}$, with $p(s)=(p^1(s),\cdots,p^l(s))\in\mathbb{R}_{++}^l$.

There are n financial assets traded in period 0. Denote the quantity vector of stocks purchased by consumer i, $z_i=(z_i(1),\cdots,z_i(j),\cdots,z_i(n))\in\mathbb{R}_+^n$, and denote $\sum_{i=1}^m z_i = z$, with associated spot price system

$q=(q(1),\cdots,q(j),\cdots,q(n))\in\mathbb{R}_{++}^n$. We assume $l(S+1)$ complete commodity markets and model producers'

sequential optimization behavior in an incomplete financial markets environment. Incomplete markets is shown to be a consequence of the technological uncertainty hypothesis. Denote producer j's long run net activity vector

$y_j=(y_j^m(s)\times y_j^n(s),\cdots,y_j^m(S)\times y_j^n(S))\in\mathbb{R}^{lS}$,

where $y_j^m(s)\in\mathbb{R}_-^m$ represents the long run input vector and $y_j^n(s)\in\mathbb{R}_+^n$ the associated feasible output vector. A state s net activity of the firm j is denoted $y_j(s)=(y_j^1(s),\cdots,y_j^l(s))\in\mathbb{R}^l$, where by convention an element $y_j^k<0$ denotes a factor of production and $y_j^k\geq 0$ a good produced. Let $\sum_{j=1}^n y_j = y$ denote the long run net activity vectors.

Sequential behavior of the producers: Consider the sequential structure of the optimization problem of the firm. Firms build up long run production capacity in the first period, for that, they issue stocks. The value of

total stocks issued in period one, denoted $qz_j=m_j$, where $m_j\in\mathbb{R}$ is a real number, bounds the quantity of goods a producer j can buy in state $s\in S$ at input prices $p_{INP}(s)$ in period two. Once money is received

through financial markets, firms install production capacity, and production activities take place subject to constraint long run production sets in the second period. Uncertainty in production is introduced by a random variable $s\in S_j$ for every j. We assume that there are less uncertain states of the world S than financial assets n available for wealth transfer. Hence $n<S$ is out default asumption.

Assumption (T):

For every production set $Y_j(s)$, $s \in S_j \geq 2.$

Assumption (P):

Firms maximize long run profits.

Assumptions (F):

(i) For each j, $Y_j|_z \subset \mathbb{R}^{lS}$ is closed, convex, and $\left(\omega + \sum_{j=1}^{n} Y_j|_z\right) \cap \square_{+}^{lS}$ compact $\forall \omega_i \in \mathbb{R}_{++}^{lS}$. $0 \in Y_j|_z \Leftarrow Y_j|_z \supset \mathbb{R}_{-}^{lS}$. $Y_j|_z \cap \square_{+}^{lS} = \{0\}$. (ii) For each j, denote $\partial Y_j|_z \subset \mathbb{R}^{nS}$ a C^{∞} manifold for transformation maps (1) $\phi_j : \mathbb{R}_{-}^{m} \times \mathbb{R}_{-}^{n} \to \mathbb{R}_{+}^{l}$ non-linear for all $s \in S$ 1.

Production takes place in the second period, once capacity is installed and state $s \in S$ occurred. At $t = 0,$ firms choose z_j at price q such that long run profits are maximized in every state $s \in S$ subject to long run technological feasibility ϕ_j and capacity constraints m_j. Denote the long run production set $Y_j|_z$. This set is

not independent of the firm's technology nor on its financial activities, denoted Z. More formally, the firm's sequential optimization problem is

$$(\bar{z}, \bar{y})_j \ \arg\max \left\{ \bar{p}(s)\square y_j(s) : \begin{vmatrix} y_j \in Y_j|_z \\ \bar{q} z_j = \left[p(s) \cdot y_j(s)\right]_{INP} \ \forall s \in S \end{vmatrix} \right\}. \quad (1)$$

Denote a long run equilibrium output vector associated with the production set boundary $\bar{y}_j \in \partial Y_{j,eff}|_z$. Each

firm j is characterized by set of assumptions F (Debreu [3]). We modify Debreu's assumptions on production sets in order to allow the modeling of endogenous production capacity via financial assets. The $t = 1$ maps implied by equation (1), $\pi_j : \mathbb{R}_{++}^{l} \times \mathbb{R}^{l} \times \mathbb{R}_{+} \to \mathbb{R}_{+}$, for each state $s \in S$ and all producers j define the $(S \times n)$ total long run payoff matrix, a collection of n vectors denoted

$$\Pi\left(p_1, \phi|_Z\right) = \begin{bmatrix} p(s)\cdot y_1(s) & \cdots & p(s)\cdot y_n(s) \\ \vdots & & \vdots \\ p(S)\cdot y_1(S) & \cdots & p(S)\cdot y_n(S) \end{bmatrix}, \tag{2}$$

where $\phi|_Z$ denotes the technology and capacity dependency of the payoff structure. We next introduce the consumer side of the economy.

The consumer: Each consumer $i \in \{1, \cdots, m\}$ is characterized by set of assumptions C of smooth economies (Debreu [10]).

Assumptions (C): a) $u_i : \mathbb{R}_+^{l(S+1)} \to \mathbb{R}$ is continuous on $\mathbb{R}_+^{l(S+1)}$, and C^∞ on $\mathbb{R}_{++}^{l(S+1)}$. $u_i(x_i) = \{x_i' \in \mathbb{R}_+^{l(S+1)} : u_i(x_i') \geq u_i(x_i)\} \subset \mathbb{R}_{++}^{l(S+1)}$, $\forall x_i \in \mathbb{R}_{++}^{l(S+1)}$. For each $x_i \in \mathbb{R}_{++}^{l(S+1)}$, $Du_i(x_i) \in \mathbb{R}_{++}^{l(S+1)}$, $\forall s$. For each $x_i \in \mathbb{R}_{++}^{l(S+1)}$, $h^{\mathrm{T}} D^2 u_i(x_i) h < 0$, for all nonzero hyperplane h such that $(Du_i(x_i))^{\mathrm{T}} h = 0.$ b) Each i is endowed with $\omega_i \in \mathbb{R}_{++}^{l(S+1)}$.

Consumers want to transfer wealth between future spot markets. For that, they invest in firms in period $t = 0$, receiving a share of total dividend payoffs which are determined in the next period in return. Denote the sequence of $(S+1)$ budget constraints

$$B_{z_i} = \left\{ x_i \in \mathbb{R}_{++}^{l(S+1)}, z_i \in \mathbb{R}_+^n : \begin{array}{l} p(0)\cdot(x_i(0) - \omega_i(0)) = -qz_i \\ p(s)\square(x_i(s) - \omega_i(s)) = \Pi(p_1, \phi)\theta(z_i) \end{array} \right\}, \tag{3}$$

where[2] ownership structure is a $(n \times 1)$ vector defined by the mappings

$$\theta_{ij} : \mathbb{R}_+ \to \mathbb{R}_+, \forall j, \tag{4}$$

where $z_i(j) \in \mathbb{R}_+$ is a positive real number for every $j = 1, \cdots, n.$ $\theta_{ij} = z_i(j)\left[\sum_i z_i(j)\right]^{-1}$ is the proportion of total payoff of financial asset j hold by consumer $i \in I.$ In compressed notation, we write

$$B_{z_i} = \left\{x_i \in \square_{++}^{l(S+1)}, z_i \in \square_{+}^{n} : p(s)\square\left(x_i(s) - \omega_i(s)\right) \in \hat{\Pi}\left[z_i \mid \theta(z_i)\right]\right\} \quad (5)$$

where $\hat{\Pi}(p_1, q, y) = \begin{bmatrix} -q_1 & \cdots & -q_n \\ p(1)\cdot y_1(1) & \cdots & p(1)\cdot y_n(1) \\ \vdots & & \vdots \\ p(S)\cdot y_1(S) & \cdots & p_1(S)\cdot y_n(S) \end{bmatrix}$ represents the full payoff matrix of order $((S+1)\times n)$.

We introduce following prize normalization $\mathfrak{S} = \left\{p \in \mathbb{R}_{++}^{l(S+1)} : \|p\| = \Delta\right\}$ such that the Euclidean norm vector of the spot price system $\|p\|$ is a strictly positive real number $\Delta \in \mathbb{R}_{++}$.

Definition 1. A financial markets equilibrium with production $(\bar{x}, \bar{y}, \bar{z}), (\bar{p}, \bar{q}) \in \mathbb{R}_{++}^{l(S+1)m} \times \mathbb{R}_{+}^{l(S+1)n} \times \mathbb{R}^{nm} \times \mathfrak{S} \times \mathbb{R}_{++}^{n}$ sat isfies:

a) $(\bar{x}_i; \bar{z}_i) \arg\max\left\{u_i(x_i; z_i) : \bar{x}_i \in B_{\bar{z}_i}(\bar{p}, \bar{q}, \bar{y}; \omega_i)\right\} \quad \forall i.$

b) $(\bar{z}, \bar{y})_j \arg\max\left\{\bar{p}(s)\square y_j(s) : \begin{vmatrix} y_j \in Y_j|_z \\ \bar{q}z_j = \left[\bar{p}(s)\cdot y_j(s)\right]_{INP} \quad \forall s \in S \end{vmatrix}\right\} \quad \forall j.$

c) $\sum_i^m (\bar{x}_i - \omega_i) = \sum_j^n \bar{y}.$

d) $\sum_i^m \theta(\bar{z}_i)_j = 1 \quad \forall j, \text{ and } \sum_{j=1}^n \sum_{i=1}^m (\bar{z}_i)_j = 0.$

a) and b) are the optimization problems of the consumers and producers. c) and d) represent physical goods and financial markets clearance conditions. $\forall j \sum_{i=1}^{m} \theta(\bar{z}_i)_j = 1$ states that each firm j is owned by the consumers. We now show that incomplete markets is a consequence of technological uncertainty and then move to the main section of the paper.

Proposition 1 $n < S \Leftarrow Y_j|_{\bar{z}}$ for all j, and $S_j \geq 2.$

Proof. Let $S_j = 1$ for every j. and $\sum_j S_j = S.$ Then long run profit prospects $\pi(p) > 0$ imply long run capacity adjustment and market entrance until $n = S$. Let $S > 1$ for every j, and $\sum_j S_j = S.$ Then, $\pi(p) > 0$ implies market entrance and the issue of new securities such that in the limit as $\pi(p) \to 0$ the number of firms $j \to n < S$ by assumption (T). Similar for negative long run profit prospects $\pi(p) < 0,$ firms exit the market and $j > n \to n < S.$ □

3. GENERIC EXISTENCE FOR CONVEX SMOOTH PRODUCTION MANIFOLDS

In this section, we show existence of equilibria. The strategy of the proof is to show that a pseudo equilibrium exists and that every pseudo equilibrium is also a financial markets equilibrium with production. It is known that pseudo equilibria exists for exchange economies. See Duffie, Shafer, Geanokopolos, Hirsh, Husseini, and others [9] [11] - [16] . Genakopolos et al. [8] showed that pseuedo equilibria exist for an economy with production for the case of exogenous financial markets. At variance with their model, where the firm's problem is to solve a Nash equilibrium, we show that a pseudo equilibrium for a more general price and technology dependent asset structure, permitting the modeling of production and its finance, exists.

Definition 2. if $\exists z \in \mathbb{R}^n_{++}$ s.t.

$\hat{\Pi}(p_1,q,\phi)\left[z \mid \sum_{i=1}^{m}\theta(z_i)_{s=1}^{S}\right] \geq 0,$ then $q \in \mathbb{R}_{++}^{n}$ is a no-arbitrage asset price relative to p_1.

Lemma 1. $\exists \beta \in \mathbb{R}_{++}^{S}$ s.t. $q = \sum_{s=1}^{S} \beta \square \Pi(p_1,\phi)$.

Proof. Immediate consequence of the separation theorem for $((S+1)\times n)$ matrices in Gale (1960). It asserts that either $\exists z \in \mathbb{R}_{++}^{n}$ such that $\hat{\Pi}z \geq 0,$ or $\exists \beta \in \mathbb{R}_{++}^{S+1}$ such that $\beta\hat{\Pi} = 0.$ □

We can now rescale equilibrium prices without affecting equilibrium allocations, let $P_1 = \beta \square \bar{p}_1$. The next step is to derive a normalized no arbitrage equilibrium definition [17] . Let $\beta \in \mathbb{R}_{++}^{S}$ be $\left(\frac{\lambda(s)}{\lambda}\right)_{i=1}$ the gradient vector from the optimization problem of agent 1, called the Arrow-Debreu agent. The Walrasian budget set for the Arrow-Debreu agent is a sequence of constraints denoted

$$B_1 = \left\{ x_1 \in \square_{++}^{l(S+1)} : \begin{matrix} P \cdot (x_i - \tilde{\omega}_i) = 0 \\ P(s)\square(x_i(s) - \omega_i(s)) = \sum_j \theta_{ij} P(s)\square y_j(s) \end{matrix} \right\}. \quad (6)$$

For all consumers $i \geq 2,$ the no arbitrage budget set consisting of a sequence of $(S+1)$ constraints is denoted

$$B_{i\geq 2} = \left\{ x_i \in \mathbb{R}_{++}^{l(S+1)} : \begin{matrix} P \cdot (x_i - \tilde{\omega}_i) = 0 \\ P(s)\square(x_i(s) - \omega_i(s)) \in \langle \Pi(P_1,\phi) \rangle \end{matrix} \right\}, \quad (7)$$

where $\langle \Pi(P_1,\phi) \rangle$ is the span of the income transfer space of period one. Replace $\langle \Pi(P_1,\phi) \rangle$ with L in $G^n(\mathbb{R}^S),$ where $G^n(\mathbb{R}^S)$ is the Grassmann manifold[3] with its known smooth $(S-n)n$ dimensional structure, and L an n-dimensional affine subspace of $G^n(\mathbb{R}^S)$.

Denote the pseudo opportunity set $B_i(P,L;\omega_i)$, for each i,

$$B_i = \left\{ x_i \in \mathbb{R}_{++}^{l(S+1)} : \begin{matrix} P\cdot(x_i - \tilde{\omega}_i) = 0 \\ P(s)\square(x_i(s) - \omega_i(s)) \subset L \end{matrix} \right\}. \tag{8}$$

Let $\mathfrak{S}' = \left\{ p \in \mathbb{R}_{++}^{l(S+1)} : p^{0,1} = \Delta \right\}$ be the set of normalized prices, and let $\Delta \in \mathbb{R}_{++}$ be a fixed strictly positive real number. This convenient normalization singles out the first good at the spot $s=0$ as the numeraire. We introduce following definitions for the long run payoff maps associated with sets $\mathfrak{S}$ and $\mathfrak{S}'$:

Definition 3. For any $p_1 \in \mathfrak{S}$, such that $\pi : \mathfrak{S} \times \mathbb{R}^l \times \mathbb{R}_+ \to \mathcal{A}$, let $\Gamma(P_1,\phi) = \beta\square\left[\left(proj_\Delta\left(\frac{1}{\beta}\right)^T \square P_1\right)\square y\right]$, where T denotes the transpose, $\frac{1}{\beta} = \left(\frac{1}{\beta(1)},\cdots,\frac{1}{\beta(S)}\right) \in R_{++}^S$, $proj_\Delta(z) = \Delta\left(\frac{z}{\|z\|}\right)$, and $\beta = (\beta(1),\cdots,\beta(S)) \in R_{++}^S$. (ii) For any $P_1 \in \mathfrak{S}'$, such that $\pi : \mathfrak{S}' \times \mathbb{R}^l \times \mathbb{R}_+ \to \mathcal{A}$, let $\Gamma(P_1,\phi) = \beta\square\left[\left(\left(\frac{1}{\beta}\right)^T \square P_1\right)\square y\right]$, where $\mathcal{A}$ is a set of $(S\times n)$ matrices A of order $(S\times n)$.

We can now define the pseudo financial markets equilibrium with production. We then state the relational propositions between a full rank FE with production and a pseudo FE with production.

Definition 4. A pseudo financial markets equilibrium with production $(\bar{x},\bar{y}),(\bar{P},\bar{L}) \in \mathbb{R}_{++}^{l(S+1)m} \times \mathbb{R}_{+}^{l(S+1)n} \times \mathfrak{S}' \times G^n(\mathbb{R}^S)$ satisfies:

a) $(\bar{x}_1)\arg\max\{u_1(x_1)\ s.t.\ x_1 \in B_1(\bar{P},\omega_1)\}\quad i=1.$

b) $(\bar{x}_i)\arg\max\{u_i(x_i)\ s.t.\ x_i \in B_i(\bar{P},\bar{L},\omega_i)\}\quad \forall i \geq 2.$

c) $\langle\Gamma(\bar{P}_1,\bar{\phi})\rangle \subset \bar{L}$, proper if $\langle\Gamma(\bar{P}_1,\bar{\phi})\rangle = \bar{L}.$

e) $(\bar{y})_j \arg\max\left\{\bar{p}(s)\square y_j(s) : \begin{vmatrix} y_j \in Y_j\big|_z \\ \bar{m}_j = [\bar{p}(s)\cdot y_j(s)]_{INP}\quad \forall s \in S \end{vmatrix}\right\}\quad \forall j.$

e) $\bar{x}_1 + \sum_{i=2}^{m}\bar{x}_i = \sum_{i=1}^{m}\omega_i + \sum_{j=1}^{n}\bar{y}_j.$

Lemma 2. Under assumptions C, demand mappings $f_1(P,w_1)$ and $f_i(P,L,\omega)$ for $i=2,\cdots,m$, from argmax a) and b) are C^∞. Under assumptions F, supply mappings $g_j(P)$ for $j=1,\cdots,n,$ from argmax d) are C^∞.

Proof. The details of this known result are omitted [11]. However, note that smoothness of demand and supply functions follows from the setup of the model for smooth economies. □

Proposition 2. For every full rank FE with production $(\bar{x},\bar{y},\bar{z}),(\bar{p},\bar{q}),$ there exists $\beta \in \mathbb{R}_{++}^S$ and a n-dimensional subspace $L \in G^n(\mathbb{R}^S)$ such that $(\bar{x},\bar{y}),(\bar{P},\bar{L})$ is a pseudo FE with production.

Proof. By lemma 1, there exists $\beta \in \mathbb{R}_{++}^S$ such that (FE) spot prices at $\bar{p}$ can be rescaled such that $P = \beta\square\bar{p},$ then $(\bar{x},\bar{y},\bar{z}),(\bar{p},\bar{q})$ is a $(\bar{x},\bar{y}),(\bar{P},L)$ equilibrium. Since by definition $\beta \in \mathbb{R}_{++}^S$ is $\left(\frac{\lambda^s}{\lambda^0}\right)_{i=1}$ of agent 1 at $(\bar{x},\bar{y},\bar{z}),(\bar{p},\bar{q}),$ agent 1's consumption bundle is $\bar{x}_1,$ since $\nabla u_1(x_1) = P,$ and $P\square(x-\omega)=0$.

On the contrary, if have a $(\bar{x},\bar{y}),(\bar{P},L)$ equilibrium, and $\bar{z}_2,\cdots,\bar{z}_m$ such that a) $\sum_{i=1}^{m}(\bar{x}_i-\omega_i)=\sum_{j=1}^{n}\bar{y}$, b) $\{\bar{x}_1\in\mathbb{R}_{++}^{l(S+1)}:P(x-\omega)=0\}$, c) $(\bar{x},\bar{y},\bar{z})$ solves $i\geq 2$ maximization problem for constraints $B_{z_i}^{FM}$. Then by defining $\bar{z}_1=-\sum_{i=2}^{m}\bar{z}_i$, every $(\bar{x},\bar{y}),(\bar{P},L)$ is a $(\bar{x},\bar{y},\bar{z}),(\bar{p},\bar{q})$ equilibrium.

Remark: Since agent 1 faces only the Arrow-Debreu constraints, his behavior is identical in both models.

Observation (2): Suppose $(P,\tilde{L})\in\Psi^{\rho}$ are elements of the (FE) pseudo equilibrium manifold, and conditions a) $[I \quad F]\hat{V}(P_1,\phi)=0$, and (ii) $\begin{bmatrix} Q_{\rho\times(S-n)} \\ F^{Q}_{(n-\rho)\times(S-n)} \end{bmatrix}_{n\times(S-n)}-[L]_{n\times(S-n)}=0$ hold.

Under these conditions, a consumption bundle $\bar{x}_i \ (i\geq 2)$ is feasible under the constraints b) in the ψ model if and only if $\bar{x}_i \ (\forall i)$ is feasible under the constraints holding with equality in a) in the (FE) model.

The next step is then to show that $\tilde{L}=\left(L^{\perp}\subset\left\langle\Gamma^{\rho}(P_1,\phi)\right\rangle^{\perp}\right)$ exists. Recall that

$$\Psi^{\rho}=\left\{\left(P_1,\left\langle\Gamma^{\rho}(P_1,\phi)\right\rangle^{\perp},L^{\perp}\right)\in P^{\rho}\times G^{S-n+\rho}(\mathbb{R}^S)\times G^{S-n}(\mathbb{R}^S):L^{\perp}\subset\left\langle\Gamma^{\rho}(P_1,\phi)\right\rangle^{\perp}\right\}.$$

Let $e=\left\langle\Gamma^{\rho}(P_1,\phi)\right\rangle^{\perp}$ and $l=L^{\perp}\subset\left\langle\Gamma^{\rho}(P_1,\phi)\right\rangle^{\perp}$. Relabel an element $(\hat{P},\hat{e},\hat{l})$ of Ψ^{ρ} in the orthogonal basis of $\mathbb{R}^S$ such that in the neighborhood of $\hat{e}$, the vector space e is spanned by the columns of a $S\times(S-n+\rho)$ matrix $\begin{bmatrix} 1_{(S-n+\rho)\times(S-n+\rho)} \\ E_{(n-\rho)\times(S-n+\rho)} \end{bmatrix}$. Similarly, in the neighborhood of $\hat{l}$, the vector space l in the same orthogonal basis of $\mathbb{R}^S$ is spanned by the columns of a $S\times(S-n)$ matrix $\begin{bmatrix} 1_{(s-n)\times(s-n)} \\ L_{n\times(s-n)} \end{bmatrix}$. We also rewrite the financial return matrix $V(.,.)$ in this basis, such that it becomes $\hat{V}(P_1,\phi)=\begin{bmatrix} {}_{(n-\rho)\times(n-\rho)} \\ {}_{(S-n+\rho)\times(n-\rho)} \end{bmatrix}, S\times(n-\rho)$.

Condition (1): $e=\left(span\left(\Gamma^{\rho}\left(P_1,\phi\right)\right)\right)^{\perp}$.

Translate $\begin{bmatrix} I_{(S-n+\rho)\times(S-n+\rho)} \\ E_{(n-\rho)\times(S-n+\rho)} \end{bmatrix}_{S\times(S-n+\rho)} \rightarrow [I \quad E]$ then condition (1) becomes

$$[I \quad E]\hat{V}(P_1,\phi)=0. \tag{9}$$

Condition (2): $l\in G^{S-n}\left(\mathbb{R}^S\right)\subset e\in G^{S-n+\rho}\left(\mathbb{R}^S\right)$.

Need to find a matrix Q such that $\begin{bmatrix} I \\ E \end{bmatrix}-\begin{bmatrix} I \\ L \end{bmatrix}=0.$ We first partition $\begin{bmatrix} I \\ E \end{bmatrix}$ such that it becomes $\begin{bmatrix} 1_{(S-n)\times(S-n)} & 0_{(S-n)\times\rho} \\ 0_{\rho\times(S-n)} & 1_{\rho\times\rho} \\ E'_{(n-\rho)\times(S-n)} & E''_{(n-\rho)\times\rho} \end{bmatrix}$, then

$$\begin{bmatrix} 1_{(S-n)\times(S-n)} & 0_{(S-n)\times\rho} \\ 0_{\rho\times(S-n)} & 1_{\rho\times\rho}Q_{\rho\times(S-n)} \\ E'_{(n-\rho)\times(S-n)} & E''_{(n-\rho)\times\rho}Q_{\rho\times(S-n)} \end{bmatrix} = \begin{bmatrix} 1_{(S-n)\times(S-n)} \\ 1_{\rho\times\rho}Q_{\rho\times(S-n)} \\ E'_{(n-\rho)\times(S-n)} + E''_{(n-\rho)\times\rho}Q_{\rho\times(S-n)} \end{bmatrix} = \begin{bmatrix} 1_{(S-n)\times(S-n)} \\ Q_{\rho\times(S-n)} \\ E^{Q}_{(n-\rho)\times(S-n)} \end{bmatrix}.$$

Q is a $(\rho \times S - n)$ matrix. Condition (2) can then be written in terms of Q and E:

$$\begin{bmatrix} Q_{\rho\times(S-n)} \\ E^{Q}_{(n-\rho)\times(S-n)} \end{bmatrix}_{n\times(S-n)} - [L]_{n\times(S-n)} = 0. \tag{10}$$

The final step is then to show that the pseudo equilibrium manifold Ψ^{ρ}, parameterized by P and Q is locally identified by a diffeomorphism $\Lambda(P,\tilde{L})$, defined by $(9)\times(9)\mapsto\Psi^{\rho}$. The partial derivative $D^{-1}_{P,Q}\Lambda(P,\tilde{L}(Q))$ exists, moreover, the map is bijective. □

Proposition 3. If $(\overline{x},\overline{y}),(\overline{P},\overline{L})$ is a pseudo FE with production then for every $\beta \in \mathbb{R}^{S}_{++}$, there exist financial asset prices $\overline{q} \in \mathbb{R}^{n}_{++}$ and investment portfolios $\overline{z} = (z(1),\cdots,z(n)) \in \mathbb{R}^{n}_{++}$ such that $(\overline{x},\overline{y},\overline{z}),(\overline{p},\overline{q})$ is a $(\overline{x},\overline{y})$ allocational equivalent FE with production.

Proof. Using (Definition 3), let $\overline{q} = \sum_{s=1}^{S}\left(\Gamma\left(\overline{P}_1,\overline{\phi}\right)\right)$, let $\overline{p}_1 = proj\left(\left(\frac{1}{\beta(s)}\right)^{T} \square \overline{P}_1(s)\right)$, and let $z_1 = \sum_{i=2}^{m} z_i$. The equivalence of a pseudo equilibrium with production and a financial markets with production then follows from similar arguments as in [16] . □

Long run financial payoffs depend on the technology of the firm, its production capacity installed via financial markets, and on a set of regular prices. Equilibrium does not exist for critical prices. The next step is therefore to introduce rank dependant payoff maps, and to exhibit a class of transverse price, technology, and capacity dependent maps. We will show that equilibria exists for this smooth rank dependent real asset structure, denoted π^{ρ}.

Definition 5. Define the rank dependent long run payoff maps $\pi^{\rho}:\mathbb{R}^{l}_{++}\times\mathbb{R}^{l}\times\mathbb{R}_{+}\to\mathcal{A}^{\rho}$ for $0\le\rho\le n.$ The set of reduced rank matrices A^{ρ} of order $(S\times n)$ with $rank\left(A^{\rho}\right)=(n-\rho)$ is denoted $\mathcal{A}^{\rho}$ and is of order $(S\times n)$.

Lemma 3. a) For $1\le\rho<n,$ A^{ρ} is a submanifold of A of codimension $(S-n+\rho)\rho.$ b) for $\rho=n$ the set $\mathcal{A}^{\rho}=\{\varnothing\}$ is empty, and c) for $\rho=0,$ $\mathcal{A}^{\rho}=\mathcal{A}$ the set of reduced rank matrices is equivalent to the set of full rank matrices.

Proof. Consider the open set U of $(S\times n)$ matrices $\tilde{a}=\left[\begin{array}{c|c} A_{(n-\rho)\times(n-\rho)} & B_{(n-\rho)\times\rho} \\ \hline C_{(S-n+\rho)\times(n-\rho)} & D_{(S-n+\rho)\times\rho}\end{array}\right]$ of $rank\,(\tilde{a})=(n-\rho)$ since $\det\bar{A}\neq0.$ There exists a matrix $b_{(n-\rho)\times\rho}$ such that $\begin{bmatrix}B\\D\end{bmatrix}=\begin{bmatrix}A\\C\end{bmatrix}b\Leftrightarrow b=A^{-1}B$, and $D=CA^{-1}B$ 4. □

The lemma states that, for $1\le\rho<n,$ the incomplete income transfer space is rank reduced. The rank dependent endogenized long run asset structure has following properties.

Proposition 4. a) $\pi^{\rho}\pitchfork\mathcal{A}^{\rho}$ for integers $1\le\rho\le n$. b) $\Gamma^{\rho}\pitchfork\mathcal{A}^{\rho}$ for any $\beta\in\mathbb{R}^{S}_{++}$ and integers $1\le\rho\le n.$ c) $\Gamma^{\rho}\cap\mathcal{A}$ is generic, since it is dense and open.

Proof. a) The linear map $D_{y}\pi^{\rho}$ is surjective everywhere in Y. b) This property does not change for any $\beta\in\mathbb{R}^{S}_{++}.$ c) Immediate consequence of the transversality theorem for maps. Since each set $\pitchfork\left(\Gamma^{\rho}, \mathcal{A}; \mathcal{A}^{\rho}\right)$ is residual, their intersection is residual. □

Definition 6. Denote Ψ^{ρ} the vector bundle defined by a) a basis $P^{\rho} = \left\{P \in \mathbb{R}_{++}^{l(S+1)} : \operatorname{rank}\left(\Gamma^{\rho}\left(P_1, \phi\right)\right) = (n-\rho)\right\}$, and b) orthogonal income transfer space $L^{\perp} \subset \left\langle \Gamma^{\rho}\left(P_1, \phi\right)\right\rangle^{\perp}$,

$$\Psi^{\rho} = \left\{\left(P, \left\langle \Gamma^{\rho}\left(P_1, \phi\right)\right\rangle^{\perp}, L^{\perp}\right) \in P^{\rho} \times G^{S-n+\rho}\left(\mathbb{R}^{S}\right) \times G^{S-n}\left(\mathbb{R}^{S}\right) : L^{\perp} \subset \left\langle \Gamma^{\rho}\left(P_1, \phi\right)\right\rangle^{\perp}\right\}. \tag{11}$$

We thus have defined a fiber bundle Ψ^{ρ} of codimension $l(S+1)-1-\rho^{2}$ containing the spot price system and income transfer space consisting of a base vector P^{ρ} and fiber $G^{S-n}\left(\mathbb{R}^{S-n+\rho}\right)$. We can now state the main result.

Theorem 5. There exists a pseudo FE with production $(\bar{x}, \bar{y}), (\bar{P}, \bar{L}) \in \mathbb{R}_{++}^{l(S+1)m} \times \mathbb{R}_{+}^{l(S+1)n} \times \mathfrak{S}' \times G^{n}\left(\mathbb{R}^{S}\right)$ for generic endowments. Moreover, by the relational propositions, a FE with production $(\bar{x}, \bar{y}, \bar{z}), (\bar{p}, \bar{q}) \in \mathbb{R}_{++}^{l(S+1)m} \times \mathbb{R}_{+}^{l(S+1)n} \times \mathbb{R}^{nm} \times \mathfrak{S} \times \mathbb{R}_{++}^{n}$ exists for generic endowments.

Proof. By (Proposition 4) and using (Definition 6) define an evaluation map Z^{ρ} on $\Psi^{\rho} \times \mathbb{R}_{++}^{l(S+1)m}$, where denote $\Omega = \mathbb{R}_{++}^{l(S+1)m}$ the set of the economy's total initial endowments, such that the excess demand map $Z^{\rho} : \Psi^{\rho} \times \Omega \rightarrow N.$

For the Arrow-Debreu agent have

$$Z_1^{\rho} : \Psi^{\rho} \times \Omega \rightarrow N. \tag{12}$$

The evaluation map is a submersion, since $D_{\omega_1} Z_1^{\rho} \;\; \forall \omega_1 \in \Omega$ is surjective everywhere. $\exists$ for each $\omega_1 \in \Omega$

$$Z^{\rho}_{1,\omega_1 \in \Omega} : \Psi^{\rho} \rightarrow N \pitchfork_{\omega \in \Omega_\rho} \{0\}, \tag{13}$$

where $\{0\} \subset N,$ and $\rho = 0.$ The dimension of the preimage $Z^{-1}_{1,\omega_1 \in \Omega}(\{0\})$ is $l(S+1)-1.$ By Thom's parametric transversality theorem[5], it follows that the subset $\Omega_\rho \cap \Omega$ is generic since it is open and dense. Equilibria exist. By the equivalence propositions 2 and 3 know that full rank financial markets equilibria with production exist.

For all $1 \le \rho \le n$ the preimage of the rank reduced evaluation map has dimension $l(S+1)-1-\rho^2.$ This implies that for generic endowments $\omega \in \cap_\rho (\Omega_\rho),$ for $\rho = 1, \cdots, n,$ there is no reduced rank equilibrium, since for $Z_1^{\rho}(.,\omega)$ the set of $\{0\} = \varnothing.$ □

4. CONCLUSION

This paper links the real and the financial sector in a general equilibrium model with incomplete financial markets. Production capacity available to a firm is endogenized and depends on the financial decisions of the firm in period one. At varianve to utility maximizing objective functions of firms, the model developed here considers a long run profit maximization objective function. This rehabilitates the decentralization property of the standard Arrow-Debreu model. It is shown by a parametric transversality theorem that equilibria exists.

REFERENCES

1. Arrow, K.J. (1953) Le role des valeurs boursières pour la répartition la meilleure des risques. In Economètrie. Fondements et applications de la théorie du risque en économétrie. CNRS, Paris.
2. Arrow, K.J. (1964) The Role of Securities in the Optimal Allocation of Risk-Bearing. Review of Economic Studies, 31, 91-96.
3. Debreu, G. (1959) Theory of Value. Wiley, New York.
4. Drèze, J.H. (1974) Investment under Private Ownership: Optimality, Equilibrium and Stability. in Allocation under uncertainty: Equilibrium and Optimality. In: Drèze, J.H., Ed., Wiley, New York, 129-165.
5. Grossman, S. and Hart, O. (1979) A Theory of Competitive Equilibrium in Stock Market Economies. Econometrica, 47, 293-330.
6. Duffie, D. and Shafer, W. (1986) Equilibrium and the Role of the Firm in Incomplete Markets. Research Paper 915, Graduate School of Business, Stanford University, Stanford.
7. Ekern, S. and Wilson, R. (1974) On the Theory of the Firm in an Economy with Incomplete Markets. Bell Journal of Economics and Management Science, 5, 171-180.
8. Geanokopolos, J., Magill, M., Quinzii, M. and Drèze, J. (1990) Generic Inefficiency of Stock Market Equilibrium When Markets Are Incomplete. Journal of Mathematical Economics, 19, 113-151.
9. Bottazzi, J. (1995) Existence of Equilibria with Incomplete Markets: The Case of Smooth Returns. Journal of Mathematical Economics, 24, 59-72.
10. Debreu, G. (1972) Smooth Preferences. Econometrica, 40, 603-615.
11. Duffie, D. and Shafer, W. (1985) Equilibrium in Incomplete Markets I: A Basic Model of Generic Existence. Journal of Mathematical Economics, 14, 285-300.
12. Geanokopolos, J. and Shafer, W. (1990) Solving a System of Simultaneous Equations in Economics. Journal of Mathematical Economics, 19, 69-95.
13. Hirsch, M. (1972) Differential Topology. Graduate Texts in Mathematics. Springer Verlag, New York.

14. Hirsch, M., Magill, M. and Mas-Colell, A. (1990) A Geometric Approach to a Class of Equilibrium Existence Theorems. Journal of Mathematical Economics, 19, 95-106.

15. Husseini, S.Y., Lasry, J.M. and Magill, M. (1990) Existence of Equilibrium with Incomplete Markets. Journal of Mathematical Economics, 19, 39-67.

16. Magill, M. and Shafer, W. (1990) Incomplete Markets. Chapter 30. In Hildebrand, W. and Sonnenschein, H., Eds., Handbook of Mathematical Economics, Vol. IV, 1523-1614.

17. Cass, D. (1984) Competitive Equilibria with Incomplete Financial Markets. CARESS Working Paper #84-09. University of Pennsylvania. (2006) in Journal of Mathematical Economics.

18. Dieudonnè, J. (1972) Elèments d'analyse. Gauthier-Villars, Paris.

NOTES

1Here, C^{∞} implies differentiability at any order required. The order depends on all transversality arguments employed. M denotes the inputs and n the output elements of the production set, and l = m + n.

2□ denotes the box product. A "s by s" context dependent mathematical operation. For example the s by s inner product.

3See i.e. Dieudonnè [18] for properties of the Grassmann manifold. See Duffie and Shafer for an exposition of the Grassmann manifold in economics [11].

4It is known that $\mathcal{A}^{\rho}$ constitutes a submanifold complex of $\mathcal{A}$. See Hirsch [13].

5See i.e. Hirsch for an exposition of Thom's parametric transversality theorem [13]. For more on transversality see R. Abraham and J. Robbin (1967), Transversal Mappings and Flows (W. A. Benjamin).

2

CHAPTER

On Information Creation and Its Effect in Incomplete Financial Markets

Mingren Chen

School of Finance, Shanghai Lixin University of Commerce, Shanghai, China

ABSTRACT

"Thirst for information" existing in financial markets shows that information supply is insufficient and private information is value. Agents will select to acquire economic information under the market incentive mechanism. This is the basis of information creation in incomplete financial markets. Because the information supplies insufficiency in the financial markets is exogenous, information creation is natural exogenous. From this perspective, this paper thinks that information creation has a practical base and a very important significance in financial economic field. The effect of information creation has dual character: the one is positive effect, which should improve market efficiency in long term; the other is negative effects, which should aggravate the market speculation and manipulation, cause financial bubble, and amplify financial risk in short term.

Keywords: Incomplete Financial Markets; Information Creation; Negative Effects; Positive Effects

1. INTRODUCTION

There are lots of phenomena in financial markets show that investors seem to be "thirst for information". It can be partly interpreted by investors who have asymmetric information and incomplete information. Many economists in their economic theoretical literature believe that the reason of asymmetric information or incomplete information in markets lies in the bounded rationality of agent or the transaction costs of information; they have not conducted the research from the perspective of information supply. Complete information means that any agent has complete knowledge about the economic environment. It is obvious that no one in the world can have full knowledge about their environment; it was felt that full knowledge is not necessary because the price system makes an appropriate response to all necessary information, so agent requires only information is his preference and prices. This is a very idealistic view. There are lots of phenomena shows that it is insufficient for information supply of markets.

With the same of commodity market, information supply not only refers to information suppliers are willing to offer information, but also refers to agents have the ability to buy them under a certain price. So, the information supply in a given period is not equal to the full society information stock. Information supply is restricted not only by its cost and demand, but also by development of information market itself. Generally, market mechanisms are more developed, the more information is be delivery, the lower information delivery cost, the stronger timeliness of information; on the contrary, market mechanisms are less developed, information needs must be propagated by non-market ways, so the stock information cannot be translated into effective information supply. It means that the information supply in market is insufficient. Information supply insufficient means that the information supply can not fully meet the information needs under the conditions of unlimited demands in the incomplete financial markets. Stiglitz (1993) [1] thought that the market of information product is usually inefficient. That is, the information supply on the markets is inadequate. The insufficient information supply means that the information supply cannot fully meet the information needs under the conditions of unlimited demands in the incomplete financial markets. Then agents have motivation to obtain information as possible as they can, and that means the insufficiency of information product can be improved by agents' information creation. Well then, what is information creation? What effect does it have [2]?

The next parts of this paper are as follows: the second part expounds the implication of the insufficient information supply, the third part discusses the implication of information creation, the fourth part studies the effects of information creation, and the fifth part is the conclusion of the article and further studies about this problem.

2. THE MEANS OF INSUFFICIENT INFORMATION SUPPLY

Information supply (more precisely, effective information supply) refers to information suppliers provide information to financial market in a given period and a certain price. It includes two aspects of qualification: the first is information suppliers who are willing to offer, the second is information suppliers can sell out all of their information under a certain price. Information supply is restricted not only by its cost and demand, but also by development of information market itself. Generally, market mechanisms are more developed, the more information is delivery, the lower information delivery cost, the stronger timeliness of information; on the contrary, market mechanisms are less developed, information needs must be propagated by non-market ways, so the stock information cannot be translated into effective information supply [3].

The real world is complex, and the price system itself is not so simple, for example, there exist nonlinear price (e.g. quantity discounts) and the same product in several different price; if the information structure is different, the status of economic operation are also different. Then, the "information" contained in price is not important. A well-operating economy needs more information than them contained in the price system. Therefore, in order to find the equilibrium price, agents must know the other variables other than price (information), such as the behavior of the total number of people and the market excess demand [4,5].

Then agent certainly needs the information contained in the price in order to obtain the right decision, but that was not enough, he also must know the other aspects of the economic environmental knowledge and information. So the price does include a lot of, but not all, economic information the information! Stiglitz [1] also believes that if the information has value, the market price is only an imperfect response to insiders' information. That means that agents must need the information the more the better, that is to say, the need of information is infinity. And hence, there do not really exist too much information supply, more reasonable relative to information needs, the information supply seems to be always inadequate and agent has the motivation to obtain information as possible as he can. It can be called insufficient information supply.

We can learn the problem from the following three aspects. The first aspect is agents' limited abilities of information acquisition and processing.

Information acquisition process does not just simply duplicate the external information to the people's memory, but rather a complex process about the recognition on external information's. Many studies have shown that agents often have cognitive bias in their decision-making process in the incomplete financial markets.

The second one lies in the agents' limitations of information understand ability. The process of interpreting information is the process that people give information to some kind of meaning or significance. People access the accuracy of the information is not only with the information content, but also with people who understand the information. Agents' perceptual set is also influence to the derivation of information organization category among the impact factors of investors' interpreting information. All these phenomena show that people has limited cognitive ability.

The third is determined by information stickiness in the process of information dissemination. Information stickiness refers to such a phenomenon in the economy: it is a gradually process for economic information to disseminate into economic environment, play a role and access to people. Resulting in "information gradual spread" includes information acquisition cost and the agents' re-optimization cost of decision-making. The pricing decisions based on the information are not always present information in both cases, although the price is always changing. And more, agents often interpret the information with their own attributes; the different agent has the different ways to interpret the same information. Therefore, the impact of the signal may not be homogeneous for agents and information in response to the price is a gradual process [4-6].

3. ON INFORMATION CREATION

Agents' limited abilities on acquiring, understanding, and processing information, and information stickiness, all these made information has private value.

Stiglitz (1981) [7] studied the realization forms on the private value of information. He pointed that private benefits of information produce, including created, and spread and layered information, takes two species forms: first, direct consumption effect, which makes the information gainer get a more suitable investment portfolio; the second is the market value effect, which means that if information dissemination caused the value changes of securities market, agents can get excess earnings by group-buying auction before the information dissemination, and then release the information out. Of course, if they want to achieve the market value effect of information, agents must have: 1) before the information released, they must own or have access to assets, 2) there must be incentives of information dissemination to the market value adjustment reaction information, and 3) in order to make people believe that the information being transmitted, it must be a credible information.

Because there are information acquisition costs, decision-making re-optimization cost, bounded rationality, and other restrictions during the process of information dissemination, information editing (or understanding) and information use, all these factors have resulted the information stickiness and made that agents can not be exactly same in the information utilization, methodologies, understanding and other aspects. Therefore, no matter from information providers, or from the actual needs of information are concerned, they always play an important role in the process of information generation, information dissemination and information use. This article called it as information creation that agents 1) produced new information, 2) reduced the information stickiness, increased information accuracy and speed of information dissemination, 3) re-edit the original information which resulted in a new signal in the process of information production, dissemination and use [8].

In the current financial literatures, researchers' main emphasis is the new information emergence, so it is called "Information Produce" in literatures, for example, Stiglitz, James, Diamond, Prescott, Gorton and Morrison et al. [7]. This paper argues that it must be a long process for a message to be used from its generation to final use by agents in their decision-making, due to many reasons and restrictions necessary. And in this process, agents are not in a passive "acceptance" process but in a process of personal subjective initiative to play. Individual subjective initiative plays a great role in the information used by the financial markets. The financial market participant is an "individual" in the final analysis, even if a financial institutions' decision-making is by one or several "individuals" to make. In the process of information dissemination, information stickiness made the information available on the market information is the past information more or less. If we lower the "friction" in the information dissemination process, that is, reduce the stickiness of information or reduce the costs of information acquisition or decision-making re-optimization, and increase the quality and communication speed of information. All of these will greatly enhance the efficiency of information usage; it is equivalent to increasing the information supply of financial market. Therefore, not only the generation of new information, and the dissemination and use of stock information, it is great significant for financial markets. In this sense, this article calls the process of new information generation, transmission and processing as "information creation", and not a process, in which agents only produce new information.

4. TWO EFFECTS OF INFORMATION CREATION

Information supply scarcity in the incomplete market, as well as the agents' high demand for economic information, that generally makes agent create positive information in the competitive and consummate market. When they use these information, agents will not deliberately distort information or wanton manipulation them. They should use this information in an objective manner. In such of situation, these new information, produced by information creation, reduce information stickiness, offset market information supply insufficient, and increase market prices of information effectiveness in some degree, then reduce systemic risk of financial markets; and improvement of market information effectiveness does not appear that situation pointed out by No Transactions Theorem and G-S Paradox, instead, following these new information appears, there will be produced some new market opportunities, that should prompt agents' transactions in order to optimize their Endowment allocation or their portfolio, that actions actually increase market transaction volume. The increase supply of market information makes it possible for agents to pay costs to access needful information, thus they promote their means and capacity avoiding market risks, which enhance the wealth effects of the financial markets and the effective market. This is a hi and happy effects for all, each market and the majority of market participants pray such effects appear, we call this effect as the positive effect of information creation. From the long term, the positive effects can improve market effectiveness.

However, whether from actually or theory is not able to completely foreclose the other case contrary to the positive effects, and this will make the market more inefficient. The one hand, the creator with bounded rationality can not produce information are all the positive information, and because of information stickiness and time limit of information dissemination process, it objectively cannot guarantee these positive information is still positive information in the process of dissemination to the market, for the market is always changing. From this perspective, this new information, arising from the information creative, includes some disturbance factors. These disturbance factors mainly include some errors produced in the process of information creation and bias arising in the course of the information dissemination, reception and usage. The other hand, from the practical situation of the world financial markets, whether it is mature or the developing markets, there are information manipulations and information fraud, which will generally increase the market speculation, market manipulation, financial bubbles and increase financial risks in the short term. This effect is called the negative effects of information.

In incomplete financial markets, the market competetion itself is incomplete, so competition cannot be completely excluded negative effects of information creation from the market, that is, there is market failure in information creation. What is more, because the information has private value, one can use his private information to make super profits, so agents produce negative information is a rational behavior, therefore it is not able to fully solve this problem in the market system. Whether agent creates negative information or not, depending on his trade-off between the costs and benefits of the information creation, or depending on his game between the available gains from the using these negative information and the punishment. If he thinks that he can get net income with using these negative information, he must be used them. In these aspects, it does not be established at least in theory that prompts people actively to create positive information by so-called "moral constraints", or the effect of "moral constraints" is very limited and even powerless in elimination negative information created, and then, attempt to "to de treatment city" just is a dream, or a whimsical thing. We must seek solutions outside the market in order to increase the cost of negative information creation, so as to eliminate negative information creation behavior as possible to create behavior. The program is the economic legal system. Perfecting the legal system and strict implementation them can greatly increase the costs of negative information creation and using them [7].

It is not opposing on stressed the importance of improving economic system and the market competition mechanism, but complementary, dialectical unity, parallel. Economic legal system is effectively contributed to the establishment of effective market competition mechanism, and effective market competition mechanism can promote the perfection of the economic legal system. Under the both factors combined effecting, on the one hand, we can greatly increase the cost of negative information creation, decrease the possibility of negative information creation to a minimum limit; on the other hand, although in the short term, agent can be obtained net profits by private information, but in the long run, prices will eventually make a return to its basic values.

In short, the effect of information creation has dual character: the one is that it has positive effect to improve market efficiency in long term; the other is that it has negative effects to aggravate the market speculation and manipulation, cause financial bubble, and amplify financial risk in short term.

5. CONCLUSIONS AND FURTHER RESEARCH

This paper analyzes information supply and demand in the incomplete financial markets; consider that information has private value from the perspective of information supply and demand that is people can use their advantageous position with private information to obtain excess returns. Therefore, incomplete financial markets there cannot be really exist insufficient information demand, but insufficient information supply. Because of information supply in incomplete markets, resulting in incomplete financial market prices do not fully reflect economic information; due to incomplete financial market prices can not fully reflect the economic information, which results in incomplete financial markets is non-fully effective.

From the phenomenon that the price does not fully reflect economic information, this paper discusses the reason of insufficient information supply can be attributed to three aspects, that is limited information processing abilities, the limitations of human ability to interpret information, and information stickiness. Thus, from the objective information itself, information stickiness was stronger under the influence of information acquisition costs and the cost of re-optimization decision-making and other reasons, considered the joint effect of agents' bounded rationality and other internal factors; it is these factors that cause insufficient information supply in incomplete markets.

Insufficient information supply is therefore tremendously significant to financial market. Naturally, driven by profits pursuit motive and risk-averse motive, agents are athirst for market information because of inadequate supply of information, so it has the necessity and possibility of financial markets to boost the information effective supply.

Start from the research of information supply and demand in the incomplete financial market, this article holds that, as far as information supply, because of the fact that the agent's bounded informational acquisition and understanding capability, limitations of informational treatment capability, and information stickiness, make information supply scarcity in the financial market. Since the fact that the information supply scarcity, the market price can not sufficiently react economic information; and since the price can not sufficiently react economic information, it makes the financial market is not sufficiently efficient. The intention that the agent maximize its benefits and its desire of risk aversion, that provide the agent the first incentive force to create information by producing new information, cutting down the information stickiness, increasing information accuracy, and speeding up the information propagated and its research, that is, information creation. So, the information supply scarcity in the incomplete financial market is the logical starting point of this article.

Then, what effects does information creation have? It is decided by information accuracy, information propagated velocity and the agent's information treatment capability, and so on. Very evidently, it is very different for market been in the distinct state that the effect is also distinct. The present article holds that information creation has dual character to incomplete financial market: the one is that it has positive effect to improve market efficiency in long term; the other is that it has negative effects to aggravate the market speculation and manipulation, cause financial bubble, and amplify financial risk in short term.

Firstly, agents cannot separate multi-period risk. Consequently, they regulate their portfolio in order to utilize risky securities to disperse different risk embodied in their endowment. It accelerates, in fact, market trade. After information arriving at the market, and even to known publicly at once, agents also update their belief, even if agents are perfect rationality and there is no liquidity intention. Yet, if only the market is incomplete, not only there might be greats of quantity trades, but also this is one of manifestations of rationality. So, under costly private information, whether the market is informational efficient or not, new economic information is valuable for some agents, and there exist a partially revealing rational expectation equilibrium with information creation. Hence, if market mechanism is establish on the base of valid incentive, participator might be able to create information, and further cause entire economy and market price is operating more informational efficient.

But the process of information acquisition in incomplete financial market is a tatonnement, agents cannot guarantee en post that it must be able to improve their expectation welfare, or avert all kinds of risks. Even in normally competitive market, agents with bounded rationality can create inaccurately, even wrong information, and incorrectly utilize these information, etc. Then it maybe exist lots of risks for agents to acquire and utilize these new information. If these elements come into contagion one another among agents, it can increase market risk. So in these sense, the information creation has the other probability, namely, it may intensify the market speculation, deepen the market manipulation, and promote asset pricing bubble on economy and its crush in the end, that is, enhance the risk of incomplete financial market.

Such research is only a qualitative study about information creation and its effect in incomplete financial market, there are a number of issues has not yet commenced, and there is not a research paradigm of mathematical economics, so it should still continue to deepen for these studies.

6. ACKNOWLEDGEMENTS

This work is partially supported by the item of major subject construction of Shanghai Municipal Education Commission #J51703.

REFERENCES

1. B. Greenwald and J. Stiglitz, "New and Old Keynesians," Journal of Economic Perspectives, Vol. 7, No. 1, 1993, pp. 23-44.
2. N. G. Mankiw and R. Reis, "Sticky Information versus Sticky Prices: A Proposal to Replace the New Keynesian Phillips Curve," Quarterly Journal of Economics, Vol. 117, No. 4, 2002, pp. 1295-1328.
3. E. Fama, "Efficient Capital Markets: A Review of Theory and Empirical Work," Journal of Finance, Vol. 25, No. 2, 1970, pp. 383-417.
4. S. J. Grossman, "On the Efficiency of Competitive Stock Markets Where Traders Have Diverse Information," Journal of Finance, Vol. 31, No. 2, 1976, pp. 573-585.
5. S. J. Grossman, "Further Results on the Informational Efficiency of Competitive Stock Markets," Journal of Economic Theory, Vol. 18, No. 1, 1978, pp. 81-101.
6. S. J. Grossman and J. E. Stiglitz, "On the Impossibility of Informationally Efficient Markets," American Economic Review, Vol. 70, No. 3, 1980, pp. 393-408.
7. J. E. Stiglitz, "Information and Capital Markets," NBER Working Paper 678, 1981.
8. M. Hellwig, "On the Aggregation of Information in Competitive Markets," Journal of Economic Theory, Vol. 22, No. 3, 1980, pp. 477-498.

3

CHAPTER

Large-Scale Empirical Tests of the Markov Tree Model

Harish S. Bhat [1,*] and Nitesh Kumar [2]

[1]*Applied Mathematics Unit, University of California, Merced, CA 95343, USA*

[2]*Skytree, Inc., San Jose, CA 95110, USA*

ABSTRACT

The Markov Tree model is a discrete-time option pricing model that accounts for short-term memory of the underlying asset. In this work, we compare the empirical performance of the Markov Tree model against that of the Black-Scholes model and Heston's stochastic volatility model. Leveraging a total of five years of individual equity and index option data, and using three new methods for fitting the Markov Tree model, we find that the Markov Tree model makes smaller out-of-sample hedging errors than competing models. This comparison includes versions of Markov Tree and Black-Scholes models in which volatilities are strike- and maturity-dependent. Visualizing the errors over time, we find that the Markov Tree model yields more accurate and less risky single instrument hedges than Heston's stochastic volatility model. A statistical resampling method indicates that the Markov Tree model's superior hedging performance is due to its robustness with respect to noise in option data.

Keywords: option pricing; short-term memory in asset prices; empirical performance; hedging errors

1. MOTIVATION

The Markov Tree (MT) model, introduced in our past work [1,2], is a discrete-time option pricing model that accounts for short-term memory of the underlying asset. The MT model generalizes the binomial tree model. In the binomial tree model, the underlying asset's log returns follow a random walk: at each time step, the asset flips the same biased coin to decide whether to increase or decrease.

In the MT model, the underlying asset's log returns follow a delayed random walk. There are two coins, a "+" coin and a "–" coin, each with potentially different biases and different associated increments. If the asset decreased (respectively, increased) in value at step n, then at step n+1n+1, we flip the "–" (respectively, "+") coin. The increment at step n+1n+1 thus depends on the identity of the coin that we flip at step n+1n+1, which itself depends on the result from step n. This first-order Markovian dependence on the past is what gives the MT model its name.

When we actually price an option using the MT model, we use an asymptotic approximation of the exact Markov tree price [2]. We do this because the approximation is both extremely accurate and orders of magnitude faster than computing prices with a large, discrete Markov tree. The MT asymptotic approximation formula we employ is a perturbation of the classical Black-Scholes formula [3]. We may view this perturbation as incorporating the effects of short-term memory in the underlying asset.

When researchers have sought to generalize the Black-Scholes model, a main focal point has been the assumption of constant volatility. Because volatility governs the dynamics of the underlying asset, if we assume it is constant, then it should not depend on option parameters such as strike price and expiration date. This logic is inconsistent with empirical observations such as the volatility smile. This has led to the development of stochastic volatility (SV) models, in which the volatility follows a stochastic process. Within this class of models, we focus our attention on the Heston model [4], an SV model that admits a closed-form price.

In this paper, we use market data to conduct extensive tests of the MT model. The purpose of these tests is to answer the following central question: which feature is more important to include in an option pricing model, short-term memory of the underlying asset or stochastic volatility? In our view, the true test of an option pricing model is its out-of-sample hedging performance; this view is in line with recent work that details multiple advantages of hedging tests as compared with pricing tests [5]. Therefore, to answer the central question, we compare the out-of-sample hedging performance of the MT model against the Black-Scholes and Heston models. For consistency with the literature on empirical testing of option pricing models, we also compare in-sample and out-of-sample pricing results.

The MT model is one of several published models that account for memory in the underlying asset [6,7,8,9,10,11,12,13]. Within this set of models, there are two reasons why the MT model is the simplest: (i) the model requires the estimation of only three parameters; and (ii) the MT model is fundamentally a discrete-time model, allowing us to bypass the technical difficulties of estimation and filtering for stochastic delay/functional differential equations. Increasing the complexity of the model may well enable better agreement with observed memory dynamics. To our knowledge, no option pricing model with memory has been subjected to large-scale empirical tests. It is statistically sound to begin the empirical investigation of such models by examining a relatively simple, parsimonious example, *i.e.*, the MT model.

We are aware of the deep development of SV models. There are two reasons for choosing the Heston model as a representative SV model. First, the Heston model has already been subjected to pricing and hedging tests [14,15,16,17,18,19]. Comparing MT-*versus*-Heston results with this literature, we can gauge how the MT model would perform against many other models. Second, the problem of fitting the Heston model to market data has received much more attention than the corresponding problem for other SV models. Using one of the established methods for fitting the Heston model, we are able to study the performance of this SV model on a large data set in a reasonable amount of time.

1.1. Summary of Present Work

In the present study, we use a large database of both individual equity options and index options to study the out-of-sample pricing and hedging performance of the MT, Heston, and Black-Scholes models.

We use two databases of historical option prices. The first consists of 20,050 S&P 500 index call options from 1 January 2009 to 31 December 2010. The second consists of 3,483,461 unique LIFFE Paris equity call options traded between 19 September 2009 and 18 June 2012. Using this data, we calculate and compare the out-of-sample hedging errors made by the Black-Scholes, Heston, and MT models.

In order to conduct these tests, we develop three new methods for fitting the MT model to data, a problem that we frame as a nonlinear regression problem. The first two regression methods we develop are, respectively, underconstrained and overconstrained least-squares methods. The third method is a robust regression method that uses a loss function that smoothly approximates the L1L1/absolute loss function.

1.2. Results

Our primary result is that using any of the three regression methods developed in this paper, the MT model yields better out-of-sample hedging performance than either the Black-Scholes or Heston models. This comparison includes the practitioner's version of the Black-Scholes model in which volatility is allowed to depend on option strike and maturity, as well as a version of the Black-Scholes model, often used in the empirical option pricing literature, in which volatility is independent of option strike and maturity.

For the overconstrained least-squares MT method, we analyze the regression residuals and show that they fit a generalized hyperbolic (GH) distribution with heavier-than-Gaussian tails. We then develop computational bootstrap procedures to better understand how the MT model is likely to perform in the future. Here we add noise to market option prices, either by sampling from the best-fit GH distribution or by resampling regression residuals, and we rerun our hedging tests using this simulated option data. Using both bootstrapping approaches, we compute the distribution of the difference between the absolute errors committed by the Heston and MT models. The results show that the MT model leads to more accurate and less risky hedges.

We conclude from these results that if one seeks to extend the Black-Scholes model, accounting for short-term memory of the underlying asset may be just as important as accounting for stochastic volatility.

Further insights obtained from this study are described in Section 9.

1.3. Prior Work

The literature on option pricing is vast and has been surveyed elsewhere [20,21,22]. We focus our attention on work that empirically tests the hedging performance of one or more of the models studied in this paper.

The MT and Black-Scholes models have been compared in prior work. The first test considered short- and mid-term European call options on six different stocks that were components of the CAC-40 index. Across 44 days of testing, the MT model outperformed the Black-Scholes model in out-of-sample pricing [1]. The second test considered American call options on 89 stocks that were components of the S&P 100 index. In 10 days of testing, the MT model outperformed the Black-Scholes model in aggregate out-of-sample pricing [2].

Previous work on the MT model did not explore sufficiently the different methods for fitting the model to data, instead relying on *ad hoc* procedures based on historical volatilities [1,2]. While these studies did consider genuine out-of-sample tests of pricing errors, the issue of hedging errors was left unaddressed. Prior work on the MT model compared its predictions only to those of the Black-Scholes model, leaving out comparisons to more sophisticated models such as SV models. Finally, the sample sizes of previous empirical studies of the MT model are much smaller than that of the present study.

A comparison of the hedging performance of the Black-Scholes, Heston, and Brigo-Mercurio models has been carried out [23]. The Brigo-Mercurio formula [24] is similar to the asymptotic approximation of the exact MT option price: both price a vanilla call option using a risk-neutral distribution that consists of a mixture of normals. However, the Brigo-Mercurio model differs from the MT model in one key aspect: it does not explicitly account for short-term memory of the underlying asset. Consequently, the variances of the mixture components in the Brigo-Mercurio model have no interaction with one another; each variance is a function of a distinct model parameter. In the MT model, the variances of both mixture components interact strongly with one another; each variance is a function of the same two model parameters. The Brigo-Mercurio model allows for an arbitrary number of mixture components, while the MT model allows for only two. When we restrict the Brigo-Mercurio model to two mixture components, the option price is a function of four model parameters [23] rather than three for the MT model [1].

These differences may serve to explain why the particular Brigo-Mercurio model tested in earlier work showed poorer hedging performance than either the Black-Scholes or Heston models [23]. This contrasts sharply with the results presented in this paper.

The MT model is an example of a discrete-time Markov chain option pricing model. A recent study explores a continuous-time Markov chain model, including results on numerical simulation together with estimation of parameters [25]. As is typical in this literature, [25] avoids pursuing out-of-sample hedging performance and/or a large-scale empirical test using market prices of options.

Several authors have compared the hedging performance of stochastic volatility, jump diffusion, and Black-Scholes models [5,26,27,28,29,30,31]. While this literature does include comparisons between varieties of stochastic volatility models, it is important to note that none of these papers test constant volatility models that are more complex than Black-Scholes. The question of whether a constant volatility model might outperform stochastic volatility models in a large-scale (*i.e.*, millions of options), out-of-sample hedging test has not been answered.

There does exist prior empirical work analyzing individual equity option data. One of the earliest such works [32] applies nonparametric tests to Chicago Board of Exchange (CBOE) individual equity option data to check for systematic differences between Black-Scholes and market prices. A later study [33] examines a nearly two-year span of CBOE option data on 10 stocks to test implications of the Hull-White stochastic volatility model [34].

A more recent study analyzed market data for options written on the S&P 100 index and the stocks that form its 30 largest components [35]. Using 350,000 distinct option quotes (both calls and puts), this study examines the differences between implied risk neutral distributions for index and individual equity options. The effect of variables such as the price-to-earnings ratio and market capitalization on the skewness of implied risk neutral distributions has been studied using four years of end-of-week option data for 856 unique firms [36]—this data set comprises 67,910 distinct option quotes.

Multiple studies have used individual equity option data to analyze various models for the implied volatility smile [37,38,39]—these studies each use between 14,120 and 400,000 distinct option quotes.

In this paper, we analyze a data set that is an order of magnitude larger than the largest of the data sets we have seen mentioned in the literature. Furthermore, we compute hedging errors for individual equity options, unlike all prior studies we have reviewed that use market data for such options.

2. OPTION PRICING MODELS

For a given option, let K denote the strike price, T the time in years to expiration, r the risk-free rate of interest, and S0S0 the spot price of the underlying asset. We denote the option price as a function of the form F(x,β), where x=[K,T,r,S0] and β is a vector of model parameters that must be statistically estimated from data. Typically, these model parameters include one or more volatilities.

We define moneyness as m=S0/K, and maturity as the time to expiration in days.

Let us now review three classes of models. All models that we discuss treat r as constant over the life of the option.

2.1. Black-Scholes

The underlying asset price StSt at any time t is assumed to follow the SDE

$$dS_t = \mu_{\mathrm{BS}} S_t dt + \sigma_{\mathrm{BS}} S_t dW_t, \tag{1}$$

where μBSμBS and σBSσBS are constants and WtWt is the standard Wiener process. The annualized volatility σBSσBS is assumed to be constant. The Black-Scholes European call option price is

$$F^{\mathrm{BS}}(\mathbf{x}, \beta^{\mathrm{BS}}) = \Phi(d_1) S_0 - \Phi(d_2) K \exp(-rT), \tag{2}$$

$$d_1 = \frac{\log(S_0/K) + (r + \sigma_{\mathrm{BS}}^2/2)T}{\sigma_{\mathrm{BS}}\sqrt{T}}, \qquad d_2 = d_1 - \sigma_{\mathrm{BS}}\sqrt{T}, \tag{3}$$

where Φ is the standard normal cumulative distribution function.

The Black-Scholes model is a one-parameter model with βBS=σBS.

2.2. Heston

In Heston's stochastic volatility model, the underlying asset StSt is governed by the coupled system

$$dS_t = rS_t dt + \sqrt{v_t} S_t dW_t^s, \tag{4}$$

$$dv_t = \kappa(\theta - v_t)dt + \xi\sqrt{v_t}dW_t^v, \quad (5)$$

where vtvt is the instantaneous variance of the asset price, dWst and dWvtd are Wiener processes with correlation ρdtρdt, θ is the long-term variance, κ is the rate at which vtvt reverts to θ, and ξ is the volatility of the volatility. The Heston model contains five parameters, *i.e.*, βSV=(v0,ρ,θ,κ,ξ). The closed form European call option price is [40]:

$$F^{\mathrm{SV}}(\mathbf{x}, \beta^{\mathrm{SV}}) = S_0 P_1 - K e^{-rT} P_2, \quad (6)$$

Where

$$P_1 = \frac{1}{2} + \frac{1}{\pi}\int_0^\infty \mathrm{Re}\left(\frac{e^{-i\omega \log K}\phi(\omega - i)}{i\omega\phi(-i)}\right)d\omega$$
$$P_2 = \frac{1}{2} + \frac{1}{\pi}\int_0^\infty \mathrm{Re}\left(\frac{e^{-i\omega \log K}\phi(\omega)}{i\omega}\right)d\omega$$
$$\phi(\omega) = e^A e^B e^C$$

And

$$A = i\omega(\log S_0 + rT)$$
$$B = \frac{\theta\kappa}{\xi^2}\left((\kappa - \rho\xi i\omega - d)T - 2\log\left(\frac{1 - g_2 e^{-dT}}{1 - g_2}\right)\right)$$
$$C = \frac{\frac{v_0}{\xi^2}(\kappa - \rho\xi i\omega - d)(1 - e^{-dT})}{1 - g_2 e^{-dT}}$$
$$d = \sqrt{(\rho\xi i\omega - \kappa)^2 + \xi^2(i\omega + \omega^2)}$$
$$g_2 = \frac{\kappa - \rho\xi i\omega - d}{\kappa - \rho\xi i\omega + d}.$$

2.3. Markov Tree

For k∈{−1,0,1}, let Zk be a discrete random variable that achieves the outcomes {lk,−lk} with probabilities {qk,1−qk. Then, as originally proposed [1], the MT model assumes the underlying asset price SnSn follows the persistent (or delayed) random walk

$$n = 1 : \log S_1 = \log S_0 + Z_0 \tag{7a}$$

$$n \geq 1 : \log S_{n+1} = \log S_n + \begin{cases} Z_1 & S_n \geq S_{n-1} \\ Z_{-1} & S_n < S_{n-1} \end{cases} . \tag{7b}$$

Note that this process generalizes the random walk in the binomial option pricing model (BOPM). In the BOPM, the increments of the process are independent and identically distributed (i.i.d.) samples of a fixed random variable, while in Equation (7), the increments of the process follow a first-order Markov chain.

This process generates a risk-neutral probability mass function (pmf); using asymptotic analysis in an appropriate continuous-time limit, this pmf can be approximated closely by a mixture of normal densities, resulting in the following expression for the price of a European call option [2]:

$$F^{\mathrm{MT}}(\mathbf{x}, \beta^{\mathrm{MT}})e^{rT} = q_0 S_0 \exp\left(\frac{\sigma_1^2}{2} + \mu_1\right)\Phi(x_1)$$

$$- (1 - q_0) S_0 \exp\left(\frac{\sigma_2^2}{2} + \mu_2\right)\Phi(x_2) - q_0 K \Phi(x_3) - (1 - q_0) K \Phi(x_4), \tag{8}$$

where

$$x_i = \frac{\mu_i + \sigma_i^2 + \log(S_0/K)}{\sigma_i}, \qquad x_{i+2} = \frac{\mu_i + \log(S_0/K)}{\sigma_i} \quad (9$$

for i∈{1,2} In this paper, we treat Equation (8) as the MT model's option price.

The MT model is a three-parameter model with $\beta^{\text{MT}} = (\sigma, \sigma_+, \sigma_-)$. The parameters qk, lk, σ1σ1, and σ2 that appear either in the stochastic process Equation (7) or on the right-hand sides of Equations (8) and (9) are all functions of the components of βMT see Appendix A for detailed algebraic expressions of these quantities.

3. REGRESSION

We now describe how we use market data to fit the three models described in Section 2.

Let VΘ,i denote the column vector of all option prices associated with the symbol Θ on day *i*. Each row of VΘ icorresponds to a particular option contract, and each such contract corresponds to a row vector of the form $\mathbf{x}_j^{\Theta,i} = (K, T, r, S_0)$. Let $X^{\Theta,i}$ denote the matrix obtained by stacking these row vectors vertically, *i.e.*,

$$X^{\Theta,i} = \begin{bmatrix} \text{---}\mathbf{x}_1^{\Theta,i}\text{---} \\ \text{---}\mathbf{x}_2^{\Theta,i}\text{---} \\ \vdots \\ \text{---}\mathbf{x}_\nu^{\Theta,i}\text{---} \end{bmatrix},$$

where ν=|VΘ,i| the length of VΘ,. Let 1∈Rν denote the column vector 1=(1,1,...,1)† , where † denotes transpose. Then, once we fix the symbol Θ and the day *i*, the third and fourth columns of XΘ, are, respectively, r1 and S01; this is because the spot price of the underlying asset depends only on Θ and *i*, while the risk-free rate of interest depends only on *i*.

For the data matrix XΘ,I, and for each option pricing model F(x,β) we let F(XΘ,i,β) denote the result of applying F(x,β) to each row o

$$F(X^{\Theta,i},\beta) = \begin{bmatrix} F(\mathbf{x}_1^{\Theta,i},\beta) \\ F(\mathbf{x}_2^{\Theta,i},\beta) \\ \vdots \\ F(\mathbf{x}_\nu^{\Theta,i},\beta) \end{bmatrix}.$$

We formulate the nonlinear regression problem

$$V_{\Theta,i} = F(X^{\Theta,i},\beta) + \epsilon_{\Theta,i}, \tag{10}$$

where ϵΘ,iϵΘ,i is a column vector of residuals. We now explain how special cases of Equation (10) can be used to fit—that is, determine the values of the regression coefficients *β*—for each of the option pricing models presented in Section 2.

3.1. Black-Scholes

3.1.1. Overconstrained L2L2

The Black-Scholes model is a one-parameter model with βBS=σBS. Our first method for finding βBSβBS involves using the entire vector VΘ,I and the entire data matrix *X*. We set

$$V_{\Theta,i} = F^{\mathrm{BS}}(X^{\Theta,i},\beta) + \epsilon_{\Theta,i} \tag{11a}$$

$$\beta^{\mathrm{BS}}_{\Theta,i} = \mathrm{argmin}_{\beta\in[0.05,0.95]} \frac{1}{2}\epsilon^{\dagger}_{\Theta,i}\epsilon_{\Theta,i}, \tag{11b}$$

yielding a volatility σBS that is independent of strike and time to expiration. This is a commonly used approach in prior empirical studies [27,30].

3.1.2. Practitioner's Black-Scholes

In this second method, we find a different βBSβBS for each row of $\begin{bmatrix} X^{\Theta,i} & V_{\Theta,i} \end{bmatrix}$:

$$[V_{\Theta,i}]_j = F^{\mathrm{BS}}(\mathbf{x}^{\Theta,i}_j, \sigma^{\Theta,i}_{(j)}). \tag{12}$$

Here [VΘ,i] j denotes the *j*-th row of the vector VΘ,I. There is no error term ϵ, because Equation (12) is a nonlinear scalar equation for the single unknown σΘ,i(j) i that can be solved almost to machine precision, even when we impose the box constraints 0.05≤σΘ,i(j)≤0.95.

By solving Equation (12) for each j=1,2,...,ν, we obtain volatilities σΘ,i(1),...,σ) that are functions of both the strike price *K* and time to expiration *T*. The σΘ,i(j) values obtained in this way are known as *implied volatilities*. This procedure is one of the methods most commonly used by practitioners for fitting the Black-Scholes model to market data.

To actually carry out the solution for both the overconstrained L2 method and this method, we use the R functionoptimize, which combines golden section search and interpolation [41].

3.2. Heston

Since βSVβSV is five-dimensional, by analogy with linear regression problems, one does not expect the least-squares solution of Equation

(10) to be unique unless VΘ,i contains at least 5 rows. Therefore, a commonly used technique is to take VΘ,I to be all available call option prices for symbol Θ and day *i* in a particular data set, so that the number of rows *v* is always more than 5. The net effect of this is to compute an estimate βSV that is strike- and maturity-independent:

$$V_{\Theta,i} = F^{\mathrm{SV}}(X^{\Theta,i}, \beta) + \epsilon_{\Theta,i} \tag{13a}$$

$$\beta^{\mathrm{SV}}_{\Theta,i} = \mathrm{argmin}_{\beta \in \mathcal{S}} \frac{1}{2} \epsilon^{\dagger}_{\Theta,i} \epsilon_{\Theta,i}. \tag{13b}$$

Here $\mathcal{S} = \{(v_0, \theta, \rho, \kappa, \xi) \in \mathbb{R}^5 \mid 0.05 < v_0 < 0.95, 0.05 < \theta < 0.95, -0.9 < \rho < 0.9, 1 < \kappa < 6, 0.01 < \xi < 1.11\}.$ Similar box constraints are often used in the literature [40].

The main caveat of applying this procedure lies in the way Equation (13b) is computed. The analytical gradient of the objective function—specifically FSVFSV—is not known, and numerically computed gradients are computationally expensive and inaccurate. This is because the evaluation of FSVFSV requires the numerical computation of an oscillatory integral. For these reasons, derivative-free rather than gradient-based optimization techniques are used to solve for β [40,42].

Two popular derivative-free techniques that are used to solve Equation (13b) for Heston's model are the Nelder-Mead algorithm [43,44] and differential evolution [40]. Using artificially created option data from a set of known parameters, [40] shows that differential evolution outperforms other derivative-free optimization techniques. We choose the Nelder-Mead algorithm for two reasons. First, differential evolution requires a prohibitively large number of evaluations of the objective function to achieve reasonable accuracy for a large-scale empirical test. Second, our tests on a subsample of LIFFE option data reveal that using Nelder-Mead with 500 iterations results in better convergence than differential evolution. The specific implementation of the Nelder-Mead algorithm we use is provided by the R function optim [41].

3.3. Markov Tree

We now describe three methods to fit the MT model to data. All three methods are implemented using gradient-based optimization, leveraging the smoothness of the MT model's option price with respect to the parameters βMT.

3.3.1. Overconstrained L2L2

The first method we consider is analogous to the overconstrained procedures described above for the Black-Scholes model and Heston's stochastic volatility model. Using the full set of market option quotes for symbol Θ and day *i*, we formulate the regression problem and least-squares solution as

$$V_{\Theta,i} = F^{\mathrm{MT}}(X^{\Theta,i}, \beta) + \epsilon_{\Theta,i} \tag{14a}$$

$$\beta^{\mathrm{MT}}_{\Theta,i} = \mathrm{argmin}_{\beta\in\mathcal{B}} \frac{1}{2}\epsilon^{\dagger}_{\Theta,i}\epsilon_{\Theta,i}, \tag{14b}$$

where $\mathcal{B} = \{(x^1, x^2, x^3) \in \mathbb{R}^3 \mid 0.05 \le x^j \le 0.95,\ j = 1, 2, 3\}.$

This formulation produces an MT volatility vector $\beta^{\mathrm{MT}}_{\Theta,i}$ that does not depend on option strikes and maturities.

3.3.2. Robust MT: Overconstrained Pseudo-Huber

Consider the loss function

$$L(x) = \delta\left(\sqrt{1 + \left(\frac{x}{\delta}\right)^2} - 1\right), \tag{15}$$

adapted from the pseudo-Huber loss function ([45] (p. 619)). With *δ* fixed, L(x) behaves for small *x* like the quadratic loss x2/(2δ), while for large *x*, L(x) behaves like the absolute loss |x||x|. In this way, L(x) gains the absolute loss function's robustness with respect to outliers. At the same time, L(x) respects the property that for normally distributed errors (*i.e.*, non-outliers), using the quadratic loss will yield a maximum likelihood estimate. More precisely, a Taylor series expansion of L(x) about x=0 yields

$$L(x) = \frac{x^2}{2\delta} + O(x^4).$$

Meanwhile, with δ fixed,

$$\lim_{|x|\to\infty} \frac{L(x)}{|x|} = 1.$$

With this loss function, we can formulate an alternative solution to the regression problem Equation (14a), one in which we replace the squared error loss function with L(x)L(x):

$$\beta_{\Theta,i}^{\mathrm{MT}} = \mathrm{argmin}_{\beta\in\mathcal{B}} \sum_{j=1}^{\nu} L(\epsilon_{\Theta,i;j}),$$

(16)

where $\epsilon_{\Theta,i;j}$ is the *j*-th component of the $\epsilon_{\Theta,i;j}$ vector defined as in Equation (14a). The estimate $\beta_{\Theta,i}^{\mathrm{MT}}$ computed in this way is strike- and maturity-independent.

Unlike the L1 or absolute loss function |x||x|, for all δ>0 and for all *x*, L(x) is smooth, enabling us to use gradient-based optimization to solve Equation (16). We view the resulting solution $\beta_{\Theta,i}^{\mathrm{MT}}$ as a robust solution to the regression problem Equation (14a). In our work, we set δ=0.01.

3.3.3. Practitioner's MT: Underconstrained L2

As in the practitioner's Black-Scholes method described in Section 3.1.2, we consider a special case of Equation (10) in which we use only one row of the data matrix XΘ,iXΘ,i and the corresponding row of the vector VΘ,i:

$$[V_{\Theta,i}]_j = F^{\mathrm{MT}}(\mathbf{x}_j^{\Theta,i}, \boldsymbol{\beta}_{(j)}) + \epsilon_{\Theta,i}. \tag{17}$$

Since $\boldsymbol{\beta}_{(j)}^{\mathrm{MT}}$ is three-dimensional, the problem Equation (17) is underconstrained, *i.e.*, the set $\mathcal{F}_j = \{\boldsymbol{\beta} \mid [V_{\Theta,i}]_j = F^{\mathrm{MT}}(\mathbf{x}_j^{\Theta,i}, \boldsymbol{\beta})\}$ is infinite. For this reason, we treat the nonlinear equation as a constraint, and solve the problem

$$\boldsymbol{\beta}_{\Theta,i;j}^{\mathrm{MT}} = \mathrm{argmin}_{\boldsymbol{\beta} \in \mathcal{F}_j} \|\boldsymbol{\beta} - \mathrm{b}\|_2^2. \tag{18}$$

where b=(0.5,0.5,0.5). This formulation gives a strike- and expiration-dependent set of MT volatilities. One can view this method as a practitioner's Markov Tree method.

Recall from Section 2.3 that the parameter vector βMTβMT can be used to calculate three risk-neutral probabilities Let $\{q_k\}_{k=-1}^{1}$. Let $\mathcal{Q} = \{\boldsymbol{\beta} \in \mathbb{R}^3 \mid 0 < q_k(\boldsymbol{\beta}) < 1, \; -1 \leq k \leq 1\}$. For all index options considered in this study, when we find the solution (18) and insert it into (17), the residual error ϵ is zero to machine precision. We also find that $\boldsymbol{\beta}_{\Theta,i;j}^{\mathrm{MT}} \in \mathcal{Q}$, *i.e.*, the qkqk's are valid probabilities.

For approximately 103103 individual equity options considered in this study, solving Equation (18) results in $\boldsymbol{\beta}_{\Theta,i;j}^{\mathrm{MT}} \notin \mathcal{Q}$. For only these options, we solve the following problem instead of Equation (18):

$$\boldsymbol{\beta}_{\Theta,i;j}^{\mathrm{MT}} = \mathrm{argmin}_{\boldsymbol{\beta} \in \mathcal{F}_j \cap \mathcal{Q}} \|\boldsymbol{\beta} - \mathrm{b}\|_2^2. \tag{19}$$

In practice, this produces solutions that satisfy both the Fj and Q constraints.

3.3.4. Implementation Details

To obtain either of the overconstrained solutions (14b) or (16), we use the L-BFGS-B algorithm [46]. This is a quasi-Newton solver that uses a limited memory (L) version of the BFGS update formula, while also handling box constraints (B). Our use of a quasi-Newton solver means that we avoid calculating the exact Hessian of the objective function. The L-BFGS-B implementation we use is built into the optim command in R [41].

To obtain either of the underconstrained solutions (18) or (19), we use the package nloptr [47], an R interface to the nloptpackage [48]. Specifically, we use this package's implementation of an augmented Lagrangian method [49,50].

For all of the codes/algorithms just mentioned, we pass user-defined functions that use exact formulas to compute the gradient of the MT option price FMTFMT with respect to βMTβMT. This gradient is given in detail in Appendix B.

For the overconstrained methods, our results show that at a computed optimum βMTβMT, the gradient of the objective function is near zero. We also find that either the Hessian at the optimum is positive definite, or the computed optimum lies on the boundary of the feasible set BB.

For the underconstrained methods, we rely on the optimization algorithm to single out a unique element of the feasible set. That is, we are less interested in whether the algorithm converges to a local minimum, and more interested in how well the constraints are satisfied. In all of our tests, the solution of either Equation (18) or Equation (19) yields (i) valid risk-neutral probabilities {qk}1k=−1 and (ii) residual errors in Equation (17) that are zero to at least four decimal places, sufficient for the purposes of this study.

4. TESTS

In the previous section, we described six procedures for fitting an option pricing model to market data: two Black-Scholes procedures, one stochastic volatility procedure, and three MT procedures. All six procedures can be viewed as special cases of Equation (10). When we evaluate Equation (10) at the optimized value β=βΘ,I, the resulting value of ϵΘ,iϵΘ,i will be referred to as the *in-sample pricing error*.

4.1. Out-of-Sample Pricing Error

Having computed $\beta^{\Theta,i}$, we use this vector of regression coefficients to price options on the symbol Θ on day i+1i+1. This leads to the one day *out-of-sample pricing error*

$$e_{\Theta,i+1} = V_{\Theta,i+1} - F(X^{\Theta,i+1}, \beta^{\Theta,i}). \qquad (20)$$

Note that $V_{\Theta,i+1}$ consists of all available call option prices in our data set for day i+1i+1 and symbol Θ. The data matrix $X^{\Theta,i+1}$ is such that the risk-free interest rate and spot prices are current as of day i+1, while the strikes and the expiration dates are read from the call option contracts available on day i+1 for symbol Θ.

4.2. Out-of-Sample Hedging Error

Hedging is the process of creating a risk-free portfolio consisting of risky assets. Owing to transaction costs and other financial considerations, a simple and practical form of such a portfolio is one that uses the minimum number of financial instruments [27]. Hedging with a portfolio consisting only of an option and shares of its underlying asset is commonly referred to as single instrument hedging.

Consider a portfolio created by selling *one* call option at the price $V^{K,T}_{\Theta,i}$ and buying *n* shares of its underlying asset at the price of S0S0 per share. The residual cash obtained through this transaction at time t=0 on day *i* is $\pi^{K,T}_{\Theta,i} = V^{K,T}_{\Theta,i} - nS_0.$. The value of this portfolio depends on the market price of the call option and the market price of its underlying asset. A change in the price of the underlying asset at time t=Δt leads to a change in the call option price and the value of the portfolio. We seek a portfolio whose value is insensitive to small changes in the underlying asset price. For the BS and MT models, this can be achieved by choosing *n* such that

$$\frac{\partial \pi_{\Theta,i}^{K,T}}{\partial S_0} = \frac{\partial V_{\Theta,i}^{K,T}}{\partial S_0} - n = 0.$$

Approximating the market price $V_{\Theta,i}^{K,T}$ by the model price $F(\mathbf{x}_j^{\Theta,i}, \beta^{\Theta,i-1})$, we have

$$n = \frac{\partial V_{\Theta,i}^{K,T}}{\partial S_0} \approx \frac{\partial F(\mathbf{x}_j^{\Theta,i}, \beta^{\Theta,i-1})}{\partial S_0}. \tag{21}$$

With this *n*, the portfolio described above with value πK,TΘ,I T is often called a *delta neutral* portfolio. For the MT model, we report the exact forms of *n* in Appendix C. Since the only stochastic term in the Black-Scholes and MT models is the underlying asset price, for these models, a delta neutral portfolio can be created using a single option and its underlying asset. This cannot be done for a stochastic volatility model, as both the underlying asset price and volatility are stochastic. Therefore, for Heston's model, and for a portfolio consisting of a call option and shares of its underlying asset, we choose *n*such that the variance of the portfolio is minimized [27]. This is called a minimum variance hedge.

To evaluate the hedging performance of different models, we first construct a risk-free portfolio at time t=0 (day *i*) consisting of an option (with strike *K*and time to expiration *T*) and *n* shares of its underlying asset (with symbol Θ). To calculate this *n*, we use Equation (21) with *β* computed using data on day i−1i-1. We invest the residual cash generated into a risk-free bond maturing at time $t = \Delta t \text{ (day } i+1)$. We assess the value of the stock and option at time $t = \Delta t \text{ (day } i+1)$ by closing the portfolio, thereby generating a cash value of $nS_{\Delta t} - V_{\Theta,i+1}^{K,T-\Delta t}$. Maturity of the bond generates another $\pi_{\Theta,i}^{K,T} e^{r_i \Delta t}$ on day i+1. The hedging error of this self-financed portfolio created on day *i* and liquidated on day i+1 is then

$$H_{\Theta,i}^{K,T} = \pi_{\Theta,i}^{K,T} e^{r_i \Delta t} - V_{\Theta,i}^{K,T-\Delta t} + n_{\text{model}} S_{\Delta t}. \tag{22}$$

5. DATA

5.1. LIFFE Paris Individual Equity Options

Market data on electronically traded LIFFE option contracts was obtained from LIFFE, prior to LIFFE's acquisition first by Euronext and then NYSE. To keep the analysis feasible, we consider only the subset of the data consisting of all LIFFE Paris Equity Options traded between 18 September 2009 and 18 June 2012, encompassing 707 trading days. To further reduce the size of the data set, we only consider call options traded within this period, leaving 7,361,451 unique options. We then apply standard filtering techniques to improve the fitting process for all models, and to remove the bias involved in pricing options with very low trading volume [5]. Specifically, we remove from our data set

- short-term options, *i.e.*, options with maturity strictly less than seven trading days,
- deep in-the-money options (with moneyness S0/K≥1.4), and
- deep out-of-the-money options (with moneyness S0/K≤0.8,

Table 1. For each data set (labeled in boldface), we give the number of options (integer to the left of the semicolon) and their average market price (decimal to the right of the semicolon), binned by moneyness and/or maturity. Rows show different categories of moneyness m=S0/Km=S0/K, and columns separate options by maturity (time to expiration in days). Market prices for LIFFE and SPX options are in Euros (€) and USD ($), respectively.

	<60	60–180	>180	Overall
		LIFFE118		
<0.94	281,602; 0.31	280,725; 1.05	411,781; 2.60	974,108; 1.49
0.94–0.97	98,580; 0.79	65,437; 2.10	97,078; 4.11	261,095; 2.35
0.97–1	102,438; 1.24	65,866; 2.64	98,096; 4.63	266,400; 2.84
1–1.03	99,652; 1.85	63,418; 3.23	91,491; 5.01	254,561; 3.33
1.03–1.06	92,177; 2.59	59,610; 3.96	86,991; 5.67	238,778; 4.05
>1.06	506,405; 6.46	445,070; 7.95	653,051; 8.98	1,604,526; 7.90
Overall	1,180,854; 3.38	980,126; 4.68	1,438,488; 6.08	3,599,468; 4.81

	<60	60–180	>180	Overall
		LIFFE25		
<0.94	58,022; 0.26	55,454; 1.01	78,524; 1.82	192,000; 1.11
0.94–0.97	16,829; 0.61	11,843; 1.75	16,207; 2.58	44,879; 1.62
0.97–1	17,217; 0.89	12,177; 2.12	17,116; 2.98	46,510; 1.98
1–1.03	16,897; 1.20	11,750; 2.41	16,430; 3.11	45,077; 2.21
1.03–1.06	15,815; 1.58	11,204; 2.97	15,664; 3.42	42,683; 2.62
>1.06	97,161; 3.64	89,247; 5.26	124,452; 4.80	310,860; 4.57
Overall	22,1941; 1.98	191,675; 3.30	268,393; 3.49	682,009; 2.95
		SPX09		
<0.94	816; 7.73	196; 29.49	67; 52.02	1079; 14.44
0.94–0.97	491; 20.98	110; 53.30	35; 81.42	636; 29.90
0.97–1	600; 31.65	198; 66.06	44; 97.47	842; 43.18
1–1.03	698; 43.50	184; 74.77	29; 110.95	911; 51.96
1.03–1.06	590; 57.26	131; 89.43	21; 127.96	742; 64.94
>1.06	1164; 116.77	300;. 138.94	9; 160.66	1473; 121.55
Overall	4359; 54.06	1119; 82.11	205; 87.68	5683; 60.80
		SPX10		
<0.94	3157; 2.88	1484; 9.70		4641; 5.06
0.94–0.97	1928; 8.80	293; 33.03		2221; 12.00
0.97–1	1951; 20.43	405; 50.13		2356; 25.53
1–1.03	1678; 37.94	314; 64.81		1992; 42.18
1.03–1.06	1119; 59.51	90; 87.86		1209; 61.62
>1.06	1717; 122.25	231; 146.97		1948; 125.18
Overall	11,550; 35.16	2817; 37.84		14,367; 35.68

leaving us with 3,483,461 call options in the LIFFE data set. We denote this data set by LIFFE118, since there are options on 118 distinct stocks. The number of unique options in the LIFFE118 data set is approximately 100 times greater than the number of options analyzed in prior empirical hedging studies [26,27,51].

For this data set, the option price that we use is the LIFFE settlement price. Settlement prices are determined using the trade-weighted average market value of the option together with a number of technical considerations spelled out in LIFFE guidelines [52]. Using these settlement prices avoids well-known pitfalls associated with daily closing prices [32].

We analyze two versions of the LIFFE data set. The first, LIFFE118, has been described above. The second, denoted by LIFFE25, is the subset of LIFFE data comprising options on 25 non-dividend-paying stocks. In the top half of Table 1, we bin options in the LIFFE118 and LIFFE25 data sets by moneyness and/or maturity, and we report both the number of options (integer to the left of the semicolon) and the average market price (decimal to the right of the semicolon) of all options in each bin. Market prices are given in Euros (€). Note that, for all moneyness-maturity categories, the average price of options on stocks that do not offer dividends is smaller than the average price of all options.

5.2. SPX Options

CBOE market data on traditional European style options on the S&P 500 index for 2009 and 2010 is available fromhttp://www.deltaneutral.com. Options on the S&P 500 index have been considered in empirical studies before [26,27,29] and are known to be the benchmark options to test European option pricing models [26,32]. Again, we follow standard filtering techniques [5]; after removing all put options, we further remove

- short-term options, *i.e.*, options with maturity strictly less than seven trading days,
- deep in-the-money options (with moneyness $S0/K \geq 1.3$),
- deep out-of-the money options (with moneyness $S0/K \leq 0.8$), and
- options with zero trading volume,

leaving us with an overall SPX data set made up of 5683 call options for 2009, and 14,367 call options for 2010. This data set consists of bid and ask prices; following standard convention [5,27], we take the midpoint of the bid and ask prices to be the the market option price.

For our analysis, we group SPX options by calendar year into the subsets SPX09 and SPX10. In the bottom half of Table 1, we bin options in the SPX09 and SPX10 data sets by moneyness and/or maturity, and we report both the number of options (integer to the left of the semicolon) and the average market price (decimal to the right of the semicolon) of all options in each bin. Market prices are given in USD ($). Note that, after filtering, there are no options with maturity exceeding 180 days (*i.e.*, long-term options) in 2010.

5.3. Interest Rates

We use 90-day LIBOR rates as our proxy for the risk-free rate of interest. The risk-free rate for any option contract dated in any given month is taken to be the 90-day LIBOR rate at the beginning of the corresponding month.

6. HEDGING RESULTS

In this section, we present out-of-sample hedging results on the four data sets described in Section 5: LIFFE118, LIFFE25, SPX09, and SPX10. Our primary result is that, when applied to any of these data sets, the three different versions of the MT model described in Section 3.3 all lead to smaller out-of-sample hedging errors than either Heston's model or the Black-Scholes model.

All entries of Table 2 are computed in the following way. First, we compute the out-of-sample hedging error for each option in the indicated data set. We then bin options into moneyness-maturity categories. We include "Overall" bins that denote either all options, options binned only by moneyness, or options binned only by maturity. Finally, we calculate the mean absolute error in each bin. All LIFFE errors are in Euros (€), while all SPX errors are in US dollars ($).

Note that an option must exist in our data set on three consecutive days in order to calculate the hedging errors using the practitioner's methods; for the overconstrained methods, the option must exist for two consecutive days.

6.1. Out-of-Sample Hedging Errors for Overconstrained L2 Models

In the top half of Table 2, we compare the overconstrained L2L2 Black-Scholes method from Section 3.1.1, Heston's stochastic volatility model—also fit using an overconstrained L2L2 loss function (13b), and the MT model using the overconstrained L2L2 fitting procedure from Section 3.3.1.

Table 2. For each of six models and four data sets (given in boldface), we provide out-of-sample hedging errors binned by moneyness and maturity. Rows and columns separate options by moneyness (m=S0/Km=S0/K) and maturity (time to expiration in days), respectively. Errors for LIFFE and SPX options are in Euros (€) and USD ($), respectively. The results are discussed in detail in Section 6.1 (top half) and Section 6.2(bottom half).

	Black-Scholes				Markov Tree				Heston			
	<60	60–180	>180	Overall	<60	60–180	>180	Overall	<60	60–180	>180	Overall
LIFFE118												
<0.94	0.0539	0.0554	0.0502	0.0527	0.0336	0.0391	0.0381	0.0371	0.0483	0.0623	0.0640	0.0591
0.94–0.97	0.0547	0.0518	0.0727	0.0607	0.0475	0.0484	0.0445	0.0466	0.0700	0.0836	0.0782	0.0765
0.97–1	0.0603	0.0705	0.0915	0.0744	0.0521	0.0521	0.0470	0.0502	0.0753	0.0888	0.0802	0.0805
1–1.03	0.1117	0.1016	0.1081	0.1079	0.0536	0.0528	0.0491	0.0517	0.0732	0.0889	0.0835	0.0809
1.03–1.06	0.1933	0.1444	0.1341	0.1593	0.0529	0.0557	0.0525	0.0534	0.0654	0.0897	0.0847	0.0786
>1.06	0.4178	0.3333	0.2485	0.3250	0.0378	0.0551	0.0562	0.0501	0.0366	0.0699	0.0772	0.0625
Overall	0.2278	0.1913	0.1537	0.1880	0.0415	0.0498	0.0490	0.0468	0.0510	0.0725	0.0747	0.0664
LIFFE25												
<0.94	0.0382	0.0408	0.0355	0.0378	0.0267	0.0330	0.0258	0.0281	0.0351	0.0508	0.0432	0.0430
0.94–0.97	0.0361	0.0454	0.0578	0.0465	0.0343	0.0374	0.0288	0.0331	0.0472	0.0623	0.0502	0.0523
0.97–1	0.0448	0.0628	0.0705	0.0591	0.0366	0.0392	0.0311	0.0353	0.0493	0.0632	0.0518	0.0539
1–1.03	0.0740	0.0821	0.0796	0.0782	0.0363	0.0368	0.0295	0.0339	0.0470	0.0610	0.0505	0.0520
1.03–1.06	0.1191	0.1142	0.0934	0.1083	0.0360	0.0410	0.0305	0.0353	0.0447	0.0639	0.0506	0.0520
>1.06	0.2657	0.2343	0.1485	0.2096	0.0266	0.0387	0.0318	0.0322	0.0260	0.0504	0.0464	0.0412
Overall	0.1476	0.1399	0.0980	0.1258	0.0294	0.0370	0.0296	0.0317	0.0348	0.0536	0.0466	0.0448

						SPX09						
<0.94	1.5960	1.6972	1.7254	1.6192	1.0208	1.5097	2.1006	1.1582	1.5538	3.0197	4.2792	1.9369
0.94–0.97	1.6073	1.4276	1.4424	1.5701	1.5660	1.8457	1.7572	1.6203	2.4214	3.8394	4.0458	2.7325
0.97–1	1.5668	1.4129	1.7418	1.5444	1.8276	1.7252	1.8811	1.8091	2.7939	3.4550	3.9352	2.9958
1–1.03	2.4576	1.8226	2.5395	2.3573	1.7119	1.8393	2.0883	1.7452	2.6189	3.6280	4.1940	2.8353
1.03–1.06	3.9381	2.6151	2.3725	3.7387	1.5337	1.4643	1.3571	1.5206	2.1383	2.7559	2.8632	2.2313
>1.06	6.3086	4.5227	3.2818	5.9344	1.2601	1.2305	3.4344	1.2650	1.5567	1.8636	6.1118	1.6413
Overall	2.9351	2.3265	1.8930	2.7908	1.4548	1.5846	1.9519	1.4964	2.1303	3.0328	4.0640	2.3595
						SPX10						
<0.94	1.6121	1.9310		1.7063	0.8483	1.2550		0.9684	1.1571	1.8352		1.3574
0.94–0.97	1.4957	1.5417		1.5005	1.0896	1.7307		1.1570	2.0851	3.6284		2.2474
0.97–1	1.2584	1.3563		1.2722	1.3459	1.9216		1.4272	2.7414	4.1820		2.9447
1–1.03	1.9874	1.4087		1.9101	1.4541	1.7795		1.4976	2.6213	3.6391		2.7572
1.03–1.06	3.9285	2.6833		3.8763	1.5916	1.7519		1.5983	2.3154	3.6511		2.3714
>1.06	6.5603	4.8238		6.4119	1.6083	2.7452		1.7054	1.9719	4.2644		2.1679
Overall	2.2491	1.8759		2.1850	1.2223	1.5396		1.2768	2.0465	2.7290		2.1638

	Practitioner's Black-Scholes				Robust Markov Tree				Practitioner's Markov Tree			
	<60	60–180	>180	Overall	<60	60–180	>180	Overall	<60	60–180	>180	Overall
LIFFE118												
<0.94	0.0567	0.0551	0.0515	0.0540	0.0332	0.0392	0.0392	0.0375	0.0346	0.0389	0.0360	0.0364
0.94–0.97	0.0517	0.0547	0.0736	0.0607	0.0474	0.0486	0.0455	0.0470	0.0449	0.0469	0.0420	0.0443
0.97–1	0.0676	0.0730	0.0911	0.0777	0.0522	0.0523	0.0478	0.0506	0.0483	0.0500	0.0443	0.0472
1–1.03	0.1171	0.1055	0.1093	0.1115	0.0533	0.0529	0.0497	0.0519	0.0476	0.0507	0.0457	0.0477
1.03–1.06	0.1856	0.1415	0.1323	0.1553	0.0520	0.0554	0.0527	0.0531	0.0431	0.0494	0.0455	0.0455
>1.06	0.3878	0.3175	0.2468	0.3113	0.0369	0.0542	0.0556	0.0494	0.0284	0.0405	0.0459	0.0388
Overall	0.2166	0.1853	0.1535	0.1829	0.0410	0.0495	0.0492	0.0466	0.0358	0.0423	0.0428	0.0403
LIFFE25												
<0.94	0.0386	0.0393	0.0361	0.0378	0.0264	0.0330	0.0265	0.0283	0.0265	0.0315	0.0241	0.0270
0.94–0.97	0.0357	0.0491	0.0582	0.0474	0.0342	0.0374	0.0293	0.0333	0.0318	0.0355	0.0269	0.0310
0.97–1	0.0500	0.0642	0.0693	0.0608	0.0366	0.0393	0.0316	0.0354	0.0334	0.0356	0.0271	0.0316
1–1.03	0.0793	0.0899	0.0803	0.0824	0.0361	0.0367	0.0297	0.0339	0.0336	0.0381	0.0273	0.0324

1.03–1.06	0.1161	0.1107	0.0932	0.1062	0.0355	0.0408	0.0307	0.0352	0.0309	0.0359	0.0265	0.0306
>1.06	0.2496	0.2274	0.1461	0.2021	0.0259	0.0380	0.0314	0.0316	0.0205	0.0294	0.0250	0.0248
Overall	0.1418	0.1374	0.0971	0.1230	0.0290	0.0367	0.0297	0.0315	0.0257	0.0317	0.0252	0.0272
						SPX09						
<0.94	1.6455	1.5857	1.6313	1.6359	1.0213	1.5106	2.1209	1.1599	0.9814	1.5389	2.0052	1.1174
0.94–0.97	1.4960	1.3682	1.8598	1.4973	1.5755	1.8647	1.7848	1.6323	1.6725	1.7379	1.8819	1.6918
0.97–1	1.8978	1.6420	1.7582	1.8444	1.8346	1.7370	1.8993	1.8177	1.6571	1.4953	1.5277	1.6211
1–1.03	2.4728	1.6590	2.0505	2.3487	1.7155	1.8535	2.1170	1.7514	1.5463	1.2259	1.3636	1.4969
1.03–1.06	3.7469	2.6865	2.4543	3.6108	1.5061	1.4700	1.3488	1.4974	1.2811	1.3652	1.0848	1.2846
>1.06	5.7902	4.7446	4.1013	5.6050	1.1600	1.2270	2.3251	1.1784	0.9971	1.0844	1.1629	1.0126
Overall	2.7793	2.2778	1.8740	2.6716	1.4372	1.5916	1.9423	1.4835	1.3278	1.4058	1.6783	1.3524
						SPX10						
<0.94	1.1571	1.5405		1.2605	0.8024	1.2060		0.9216	0.6159	0.9517		0.7065
0.94–0.97	1.2358	1.4164		1.2514	1.0639	1.7000		1.1308	0.9466	1.4735		0.9919
0.97–1	1.2124	1.4077		1.2360	1.3324	1.9071		1.4135	1.1796	1.5360		1.2227
1–1.03	1.7781	1.2668		1.7279	1.4537	1.7812		1.4975	1.1644	1.2645		1.1742
1.03–1.06	3.4793	2.5822		3.4506	1.6123	1.7623		1.6186	1.0635	1.2231		1.0686
>1.06	5.9467	4.0908		5.8430	1.6270	2.7796		1.7256	1.0783	1.8883		1.1236
Overall	1.8476	1.5604		1.8053	1.2055	1.5083		1.2575	0.9572	1.1610		0.9872

Across all four data sets, the MT model's hedging error in almost all moneyness-maturity categories is smaller than that of the other models. On the entire LIFFE118 data set, the MT model's overall hedging errors are 41.88% lower than that of Heston's model, and 301.7% lower than that of the Black-Scholes model. Applied to LIFFE25 data, the MT model's overall hedging errors are 41.32% and 296.84% lower than for Heston's model and the Black-Scholes model, respectively. For SPX options from 2009 (2010), the MT model's overall hedging errors are 57.68% (69.47%) and 86.5% (71.13%) smaller than what we find for Heston's model and the Black-Scholes model, respectively.

There are several insights that we obtain from these results. First, the out-of-sample pricing performance is not indicative of the out-of-sample hedging performance of an option pricing model. Second, among all the performance metrics considered, the hedging errors seem to be least affected by not accounting for the dividends in the option pricing models—this is indicated by the consistency of the hedging results across Table 2. The MT model, in particular, performs consistently well regardless of whether the option's underlying asset pays a dividend.

We note another consistent trend in the top half of Table 2. For the MT model, if we bin errors *only* by maturity, then short-term options (those with maturity less than 60 days) are hedged best; similarly, if we bin errors *only* by moneyness, then the most out-of-the-money options (those with $m<0.94$) are hedged best.

Next, we visualize and assess hedging errors in a different way. For a given stock symbol Θ and a given day *i*, we sum the*raw* hedging errors due to all different options available on day *i* on the underlying asset Θ. This gives a hedging error for symbol Θ on day *i*. We sum over Θ, and thereby obtain a time series of *market hedging errors*. As there are four data sets, we obtain four time series for each of the three models that were tested.

In Figure 1, we plot these four time series for Heston's model (blue) and the MT model (red). The time series for the Black-Scholes model is omitted, because it increases the vertical scale of the plot to an extent that we miss details in the blue and red curves. Note also that in the fourth plot showing daily SPX10 hedging errors, the horizontal axis starts at May 2010—this is because, after filtering, there were no SPX options remaining in our data set from January 2010 until this date.

Apart from showing that the MT model outperforms Heston's model, these plots also reveal how the model's hedging errors vary on a daily basis. This variation has not been plotted before, even in large-scale empirical studies.

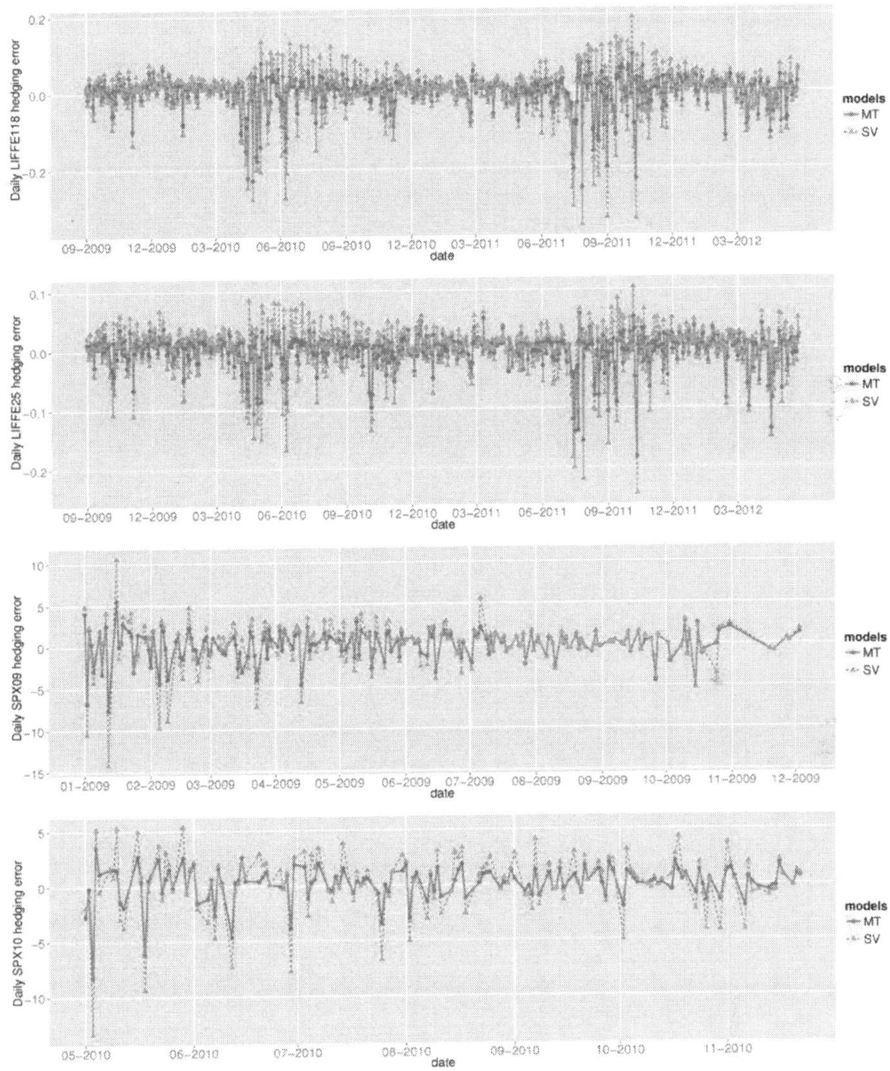

Figure 1. We plot the daily market hedging error as a function of time for Heston's model (dashed blue) and the overconstrained L2L2 Markov Tree model (solid red), applied to the four data sets LIFFE118, LIFFE 25, SPX09, and SPX10 (top to bottom). Errors for LIFFE and SPX data sets are in Euros (€) and USD ($), respectively. The results are discussed in detail in Section 6.1.

We present the following facts, which quantify trends that can be observed by examining the plots:

- The red curves are usually enveloped by the blue curves. For the four time series, the MT model's hedging errors are smaller in magnitude than those made by Heston's model with empirical probabilities between 0.760.76 and 0.850.85.
- When the blue curve has a larger error, it is usually much larger than the corresponding error for the red curve; when the red curve has a larger error, it is usually very close to the blue curve. For the LIFFE118 and LIFFE25 data sets, averaged over the days on which the MT model makes smaller-in-magnitude hedging errors, the MT model beats Heston's model by €0.01850.0185 and €0.01520.0152, respectively. If we instead average over the days on which Heston's model makes smaller-in-magnitude hedging errors, then Heston's model beats the MT model by €0.00340.0034 and €0.00470.0047, respectively. For the SPX data, taken together from 2009–2010, when the MT model wins, it wins by an average of $0.9841, while when Heston's model wins, it wins by an average of $0.2574.
- The blue curves have larger magnitude oscillations than the red curves. The standard deviations of the four MT time series are, respectively, €0.04080.0408, €0.02700.0270, $1.7462, and $1.6339. The standard deviations of the four Heston time series are, respectively, €0.06280.0628, €0.04170.0417, $2.6821, and $2.7780.

6.2. Out-of-Sample Hedging Errors for Practitioner's/Robust Models

Next, in the bottom half of Table 2, we compare the practitioner's Black-Scholes method from Section 3.1.2 against the two alternative methods for fitting the MT model: the robust method from Section 3.3.2, and the practitioner's method fromSection 3.3.3.

We see that the practitioner's MT method gives markedly better hedging results than the practitioner's Black-Scholes method. Specifically, the overall errors for the practitioner's MT method are 78%, 78%, 49% and 45% less than those of the practitioner's Black-Scholes model for, respectively, each of the four data sets. The out-of-sample pricing results did not indicate this to us, again showing the value of out-of-sample hedging tests. Note also that the practitioner's MT method performs equally well for options on dividend-paying (LIFFE118) and non-dividend-paying (LIFFE25) underlying assets.

Looking at the results in each bin, we see that for the LIFFE data sets, the practitioner's MT method outperforms the practitioner's Black-Scholes method for nearly all moneyness-maturity categories. For the SPX data set, the practitioner's MT method has far smaller errors than the practitioner's Black-Scholes method for options that are at-the-money or in-the-money. For options that are out-of-the-money, both methods yield comparable results.

Comparing these out-of-sample hedging results against those in the top half of Table 2, we reach the following conclusions:

- The overall hedging errors for the overconstrained L2L2 MT method and the robust MT method are greater than the overall hedging errors for the practitioner's MT method by 10.65% and 9.69%, respectively, for 2009 SPX options, and 29.34% and 27.38%, respectively, for 2010 SPX options. For both LIFFE data sets, the practitioner's MT method makes overall errors that are between 15.6% and 16.5% smaller than with either of the overconstrained methods. These results also show that the practitioner's MT method produces the least out-of-sample hedging errors across all data sets, not just overall, but in each individual moneyness-maturity bin. Analogous to what we found in the out-of-sample pricing results, we find that for those whose sole objective is to minimize out-of-sample hedging errors, the practitioner's MT method is the best of the three MT methods.
- On each of the four data sets, the practitioner's MT method produces substantially smaller out-of-sample hedging errors than Heston's model. This is true not only for overall errors, but also in each individual moneyness-maturity category.
- The robust MT method has smaller hedging errors than the overconstrained L2L2 MT method, in the overall category, for all four data sets.
- The robust MT method's hedging errors are smaller than those for Heston's model in nearly every moneyness-maturity category, for all four data sets. The overall errors committed by the robust MT method are 29.8%, 29.7%, 37.1% and 41.9% smaller than the overall errors committed by Heston's model for each of the four data sets, respectively.
- The overconstrained L2L2 Black-Scholes method and the practitioner's Black-Scholes method give similar hedging results. In other words, allowing the volatility to be strike- and maturity-dependent reduces the out-of-sample pricing errors, but this reduction does not carry over to out-of-sample hedging errors. This result extends earlier findings in which the Black-Scholes volatility was allowed to depend on each moneyness-maturity category [27].

7. BOOTSTRAP ANALYSIS OF HEDGING ERRORS

To better understand the superior hedging results displayed by the MT model in Section 6, we conduct a bootstrap analysis. The purpose of the bootstrap analysis is to answer the following question: if market prices of options had been different, how would the MT model have performed? The bootstrap analysis relies on statistical simulation of option prices, which we perform using generalized hyperbolic distributions that fit in-sample regression residuals.

In this section, we restrict our attention to the SPX data set, enabling us to run a reasonable number of simulations. In what follows, the underlying asset Θ will be the S&P 500 index, rather than an individual equity.

7.1. In-Sample Error Analysis

In order to statistically simulate option prices, we use our already computed regression residuals to develop error distributions. In Section 7.2, we will use these distributions in a bootstrap analysis.

While fitting the regression model as in Equation (10), we obtain the residual vector $\epsilon_{\Theta,i}$ for each day *i*. When we use the overconstrained L2L2 method from Section 3.3.1, the regression is performed under the assumption that the residuals are (i) independent of option strike and maturity; and (ii) i.i.d. samples from an error random variable. As we did previously for SPX data, we consider data from 2009 and 2010 separately.

Collecting all error vectors $\epsilon_{\Theta,i}$ for 2009 and 2010, we obtain, respectively, 5647 and 14,366 i.i.d. samples of the random variables E_{2009} and E_{2010} We first subject our samples of E_{2009} and E_{2010} to exploratory data analysis. Using expectation maximization, we then fit a generalized hyperbolic distribution (GHD) to both sets of samples. We fit a GHD for three reasons. First, the GHD has been used with success in finance where heavier-than-normal tails arise [53]. Second, the GHD includes, as special cases, other distributions of interest: hyperbolic, normal inverse Gaussian, variance gamma, student t, and normal. This property of the GHD enables us to perform likelihood ratio tests. Third, the best fit GHD is a very good match for the kernel density estimates of E_{2009} and E_{2010}—see Figure 2.

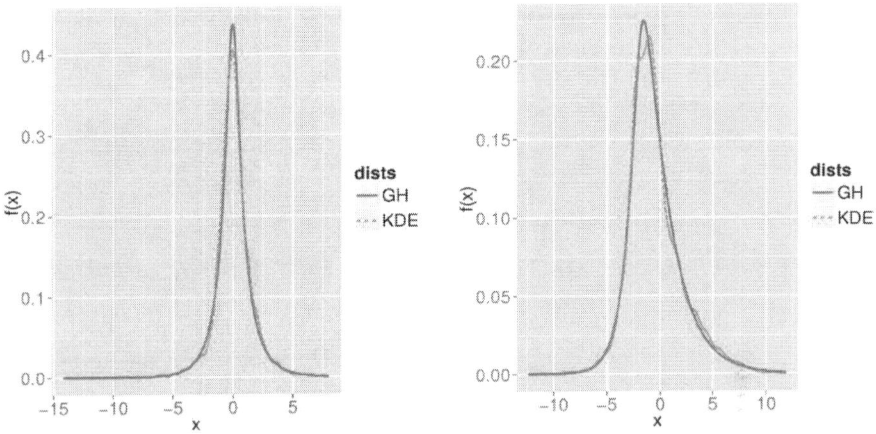

Figure 2. We plot estimated probability density functions (pdf) of the regression errors of the overconstrained L2 Markov Tree method applied to the SPX09 (left) and SPX10 (right) data sets. Solid (red) curves show kernel density estimates (KDE), and dashed (blue) curves show the generalized hyperbolic (GH) pdf using the best fit parameters from Table 3.

Using a five-dimensional parametrization ([54] (section 4.2)) of the GHD, we give the best fit parameters for E2009 and E2010E2010 in Table 3.

We employ likelihood ratio tests and AIC model selection to test the error distribution. For the samples of E2009, the likelihood ratio tests reject the hypothesis that the true underlying asset's distribution belongs to four of the five special cases of the GHD mentioned above—the exception is the normal inverse Gaussian distribution, with a *p*-value of 0.050.05. For the samples of E2010E2010, the likelihood ratio test rejects the hypothesis that the true underlying asset's distribution belongs to any of the five special cases of the GHD mentioned above. For both E2009 and E2010E2010 samples, AIC model selection criteria also selects GHD as the best model for the 2010 errors.

Taken together, these exploratory results indicate that the GHD, rather than any of the five special cases we tested, is a good fit for the residuals from 2009 and 2010. Note that the tail decay in the GHD is given by |x|aemx|x|aemx, where a=λ−1a=λ-1and *m* differs for the left and right tails. For our fitted distribution, *a* is −1.4-1.4 (0.220.22) for 2009 (2010) respectively. The exponent *m* is 0.560.56 (−0.472-0.472) for the left (right) tail for 2009 and 1.0481.048 (−0.49-0.49) for the left (right) tail for 2010. This indicates that the fitted GHD is asymmetric and has heavier tails than the normal distribution.

Table 3. Best fit generalized hyperbolic distribution (GHD) parameters for the regression errors of the overconstrained L2L2 Markov Tree method applied to the SPX09 and SPX10 data sets. For further discussion of the results of this table and Figure 2, seeSection 7.1.

	λ	$\bar{\alpha}$	μ	Σ	γ
2009	–0.4022	0.5242	–0.0803	1.4903	0.0980
2010	1.2218	0.3808	–1.8850	2.442	1.4059

The best-fit GH distributions are inconsistent with the assumption that the error ϵ in Equation (14a) is normally distributed. If ϵ does have heavier-than-normal tails, then the least-squares solution Equation (14b) will not be the maximum likelihood estimator of β in Equation (14a). This opens the door to a host of robust regression procedures for estimating β. The robust MT solution (16) we have proposed does perform marginally better than the L2L2 method in terms of overall errors. At the same time, the GHD error analysis indicates that a robust regression method tailored to the MT model may yield even better results.

7.2. MT Model Performance: Resampled Regression Coefficients

Let *j* be a fixed day in either 2009 or 2010, and let eΘ,j denote a vector of samples from the GHD using the best fit parameters from Table 3 for the appropriate year. We assume eΘ,j has the same number of components as VΘ,j in Equation (10), *i.e.*, the number of available call options for underlying asset Θ on day *j*. Let βΘ,j denote the MT regression coefficients computed using (14b) for underlying asset Θ on day *j*. Then define the vector V˜Θ,j of simulated option prices by

$$\widetilde{V}^{\Theta,j} = F^{\mathrm{MT}}(X^{\Theta,j}, \beta^{\Theta,j}) + e_{\Theta,j}. \tag{23}$$

By comparison with Equation (10), we see that if ϵ and e have the same distribution, then V˜ and V have the same distribution as well.

We now use V˜Θ,j in place of the market prices VΘ,j in Equation (14a) and obtain a new set of regression coefficients β˜Θ,j using Equation (14b). Using the coefficients β˜Θ,j on day j=i−1, we compute out-of-sample hedging errors using a self-financed portfolio created on day *i* and liquidated on day i+1i+1 as in Section 4.2. Repeating this procedure 50 times, and aggregating the data across each year, we obtain 282,350282,350 samples of HMT2009H2009MT and 718,300718,300 samples of HMT2010 H2010MT, where HMTy is a random variable representing the out-of-sample hedging error made by the MT model on one day in year *y*.

To relate the MT model's performance to that of Heston's model, we compute $\tau_y = |H_y^{\mathrm{SV}}| - |H_y^{\mathrm{MT}}|$, where HSVy is the out-of-sample hedging error for Heston's model computed as in Section 4. The reason we repeated the above procedure 50 times is so that the mean of τyτy no longer changes when further repetitions are computed.

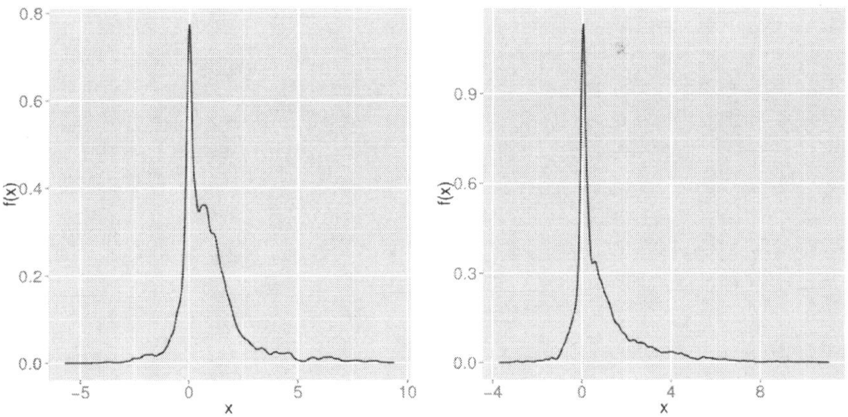

Figure 3. The random variable $\tau_y = |H_y^{\mathrm{SV}}| - |H_y^{\mathrm{MT}}|$ measures the difference in absolute out-of-sample hedging error between two models, Heston's model and the overconstrained L2 Markov Tree (MT) model, for one day in year *y*. Here we show kernel density estimates of the probability density function for τyτy for years y=2009 (**left**) and y=2010 (**right**). Positive outcomes of τyτy indicate the superiority of the MT model. Note $P(\tau_{2010} \geq 0) \approx 0.79;$ and P(τ2010≥0)≈0.79; the mean values of τ2009 and τ2010 are 0.8707 and 0.9089, respectively.

In Figure 3 we show kernel density estimates of τyτy for y=2009 and y=2010 The significant asymmetry present in both years' density plots indicate that the MT model's superior hedging performance persists even when market prices differ from historical market prices.

Note that $P(\tau_{2009} \leq 0) = 0.2366$ and $P(\tau_{2010} \leq 0) = 0.2141$, indicating that between 76%–79% of SPX options can be hedged better using the MT model. The mean values of τ2009 and τ2010 are 0.8707 and 0.9089, respectively, indicating that in repeated simulations, we expect the SV model to commit a hedging error that is between \$0.87–\$0.91 greater than that of the MT model. We report the deciles of τyτy in Table 4. The deciles show that the potential downside of the MT model is much smaller than its potential upside; for the 21%–24% of SPX options that are hedged better using Heston's model, the improvement offered by Heston's model is slight.

Finally, we note that we have run this entire resampling procedure again, but with eΘ,j denoting a vector of samples drawn uniformly with replacement from the collection of all in-sample residual error vectors ϵΘ,I *i.e.*, the vectors used above to fit the GHD distributions. Using this bootstrapping approach, when we compute the deciles of τyτy, we find values that agree to two decimal places with the values in Table 4. Unlike a perturbative analysis, the samples eΘ,jeΘ,j may be quite large, whether they are drawn from a GHD distribution with heavier-than-Gaussian tails, or drawn using the bootstrapping approach. Overall, the performance of the MT model in both types of tests indicates its robustness with respect to noisy option prices.

Table 4. We tabulate the deciles of $\tau_y = |H_y^{\mathrm{SV}}| - |H_y^{\mathrm{MT}}|$, described in Figure 3, for y=2009 and y=2010. Together with Figure 3, these results establish that the Markov Tree model yields superior performance in repeated simulations using noisy option prices. For further discussion of the results of this table and Figure 3, seeSection 7.2.

	10%	20%	30%	40%	50%	60%	70%	80%	90%
2009	−0.2888	−0.0357	0.0651	0.2879	0.5720	0.8463	1.1781	1.5894	2.4213
2010	−0.2008	−0.0066	0.0578	0.1768	0.3679	0.6658	1.0401	1.6276	2.8967

8. PRICING RESULTS

For the sake of consistency with the empirical finance literature, we include in-sample and out-of-sample pricing results for the six models and four data sets considered in this paper. Our main conclusion from this section is that neither in-sample nor out-of-sample pricing results are predictive of hedging results. Many of the empirical studies cited in Section 1may benefit from reexamination in light of this finding.

8.1. In-Sample Pricing Errors

In Table 5, we present in-sample pricing results for the three overconstrained L2L2 models. Overall, the in-sample results show that the optimization methods we have used produce reasonable values for the regression coefficients β. The general trend of the in-sample results is that, among the overconstrained methods, Heston's model does the best job of minimizing regression residuals. As Heston's model contains five parameters, two more than the MT model and four more than the Black-Scholes model, this result is not surprising.

See Table 6 for the robust MT method's in-sample errors. While the regression residuals themselves are not substantially different than those obtained from the overconstrained L2L2 MT method (see Table 5), the β's that we obtain from the robust MT method are truly different than those obtained via other procedures.

We omit from Table 6 the in-sample errors for the two practitioner's methods: in these cases, the regression residuals are zero to four decimal places, as expected.

8.2. Out-of-Sample Pricing Errors

Table 7 shows out-of-sample pricing errors for each of the four data sets described earlier. All entries of Table 7 are computed in the following way. First, we compute the out-of-sample pricing error for each option in the indicated data set. We then bin options into moneyness-maturity categories. We include "Overall" bins that denote either all options, options binned only by moneyness, or options binned only by maturity. Finally, we calculate the mean absolute error in each bin. All LIFFE errors are in Euros (€), while all SPX errors are in US dollars ($).

Table 5. For each of four data sets (in boldface), we present in-sample pricing errors for the three models studied in Section 6.1: the Black-Scholes model, the MT model, and Heston's model. All three models are fit to data using overconstrained L2L2 regression procedures. The out-of-sample pricing errors corresponding to these tables are given inTable 7. This table corresponds to the discussion of in-sample errors in Section 8.1.

	LIFFE118											
	Black-Scholes				**Markov Tree**				**Heston**			
	<60	**60–180**	**>180**	**Overall**	**<60**	**60–180**	**>180**	**Overall**	**<60**	**60–180**	**>180**	**Overall**
<0.94	0.0939	0.1767	0.3272	0.2164	0.0865	0.1484	0.2638	0.1793	0.0726	0.1263	0.2874	0.1789
0.94–0.97	0.1503	0.2204	0.3175	0.2300	0.1440	0.2114	0.2799	0.2114	0.1114	0.1358	0.2788	0.1798
0.97–1	0.1762	0.2325	0.3078	0.2386	0.1720	0.2311	0.2838	0.2278	0.1257	0.1411	0.2945	0.1917
1–1.03	0.1910	0.2563	0.3057	0.2485	0.1863	0.2533	0.2937	0.2416	0.1222	0.1473	0.3181	0.1989
1.03–1.06	0.1919	0.2810	0.3070	0.2561	0.1787	0.2664	0.2969	0.2437	0.1089	0.1547	0.3066	0.1924
>1.06	0.1440	0.3385	0.4005	0.3024	0.1236	0.2251	0.3296	0.2356	0.0631	0.1658	0.3653	0.2146
Overall	0.1431	0.2683	0.3559	0.2623	0.1302	0.2070	0.3000	0.2190	0.0834	0.1489	0.3258	0.1981
	LIFFE25											
	Black-Scholes				**Markov Tree**				**Heston**			
	<60	**60–180**	**>180**	**Overall**	**<60**	**60–180**	**>180**	**Overall**	**<60**	**60–180**	**>180**	**Overall**
<0.94	0.0684	0.1324	0.1880	0.1358	0.0634	0.1078	0.1494	0.1114	0.0421	0.0600	0.0820	0.0636
0.94–0.97	0.0936	0.1318	0.1605	0.1279	0.0903	0.1208	0.1407	0.1165	0.0567	0.0592	0.0824	0.0666
0.97–1	0.1029	0.1301	0.1624	0.1319	0.1001	0.1246	0.1497	0.1248	0.0600	0.0574	0.0896	0.0702
1–1.03	0.1035	0.1330	0.1554	0.1301	0.0999	0.1276	0.1476	0.1245	0.0559	0.0556	0.0893	0.0680
1.03–1.06	0.1054	0.1470	0.1576	0.1355	0.0971	0.1363	0.1502	0.1269	0.0500	0.0556	0.0886	0.0656
>1.06	0.0843	0.1844	0.2030	0.1605	0.0700	0.1139	0.1606	0.1189	0.0320	0.0611	0.1028	0.0687
Overall	0.0852	0.1573	0.1879	0.1459	0.0763	0.1154	0.1540	0.1179	0.0418	0.0598	0.0930	0.0670

SPX09

	Black-Scholes				Markov Tree				Heston			
	<60	60–180	>180	Overall	<60	60–180	>180	Overall	<60	60–180	>180	Overall
<0.94	0.8066	1.2891	3.3626	1.0529	0.6531	1.1272	3.1615	0.8950	0.3827	0.4417	0.4571	0.3981
0.94–0.97	1.5875	1.5894	3.5900	1.6980	1.2043	1.2083	3.2579	1.3180	0.9695	0.7099	0.7592	0.9130
0.97–1	1.4227	1.4853	3.7589	1.5595	1.1894	1.1438	3.4583	1.2972	0.9834	0.7621	1.2557	0.9456
1–1.03	1.4231	1.5424	3.1430	1.5019	1.2005	1.2169	3.1522	1.2660	0.8087	0.7603	1.3450	0.8160
1.03–1.06	1.3544	1.8294	2.2189	1.4627	1.0814	1.3002	2.4501	1.1588	0.5964	0.6742	1.5565	0.6373
>1.06	1.0156	1.3396	4.4884	1.1028	0.5414	0.9417	3.5248	0.6411	0.5892	0.4482	1.1743	0.5641
Overall	1.2080	1.4718	3.3877	1.3386	0.9048	1.1234	3.1834	1.0300	0.6838	0.6061	0.9498	0.6781

SPX10

	Black-Scholes				Markov Tree				Heston			
	<60	60–180	>180	Overall	<60	60–180	>180	Overall	<60	60–180	>180	Overall
<0.94	2.4793	4.2638		3.0499	1.8093	3.1727		2.2453	0.7678	0.7232		0.7536
0.94–0.97	2.1529	1.3254		2.0437	1.5996	1.1367		1.5385	1.9971	1.3807		1.9158
0.97–1	1.5151	3.0341		1.7762	1.3373	2.8471		1.5968	3.3062	1.2439		2.9517
1–1.03	1.9801	5.6286		2.5552	1.6103	4.7825		2.1104	3.6362	1.2564		3.2611
1.03–1.06	2.9051	8.2283		3.3014	1.7680	6.3371		2.1081	3.3663	1.2077		3.2056
>1.06	2.5464	9.5136		3.3726	1.8835	5.4483		2.3062	2.3409	1.6233		2.2558
Overall	2.2406	4.4907		2.6818	1.6727	3.3813		2.0077	2.3041	1.0152		2.0514

Table 6. For each of four data sets (in boldface), we present in-sample pricing errors for the robust MT method that utilizes the pseudo-Huber loss function. The out-of-sample pricing errors corresponding to these tables are given in Table 7 in the manuscript. This table corresponds to the discussion of in-sample errors in Section 8.1.

	LIFFE118				**LIFFE25**			
	<60	**60–180**	**>180**	**Overall**	**<60**	**60–180**	**>180**	**Overall**
<0.94	0.0735	0.1219	0.2726	0.1716	0.0545	0.0882	0.1569	0.1061
0.94–0.97	0.1245	0.1723	0.2801	0.1943	0.0775	0.0963	0.1427	0.1060
0.97–1	0.1498	0.1908	0.2851	0.2098	0.0860	0.0985	0.1567	0.1153
1–1.03	0.1634	0.2135	0.2957	0.2235	0.0858	0.1033	0.1515	0.1143
1.03–1.06	0.1581	0.2291	0.2980	0.2268	0.0840	0.1109	0.1532	0.1165
>1.06	0.1097	0.2008	0.3440	0.2303	0.0607	0.0961	0.1668	0.1133
Overall	0.1141	0.1782	0.3094	0.2096	0.0659	0.0953	0.1601	0.1112
	SPX09				**SPX10**			
	<60	**60–180**	**>180**	**Overall**	**<60**	**60–180**	**>180**	**Overall**
<0.94	0.6076	1.1512	3.7982	0.9045	1.5200	2.6941		1.8954
0.94–0.97	1.1659	1.1951	3.7680	1.3141	1.1958	1.5735		1.2456
0.97–1	1.1216	1.1430	4.1478	1.2848	1.2525	3.7508		1.6820
1–1.03	1.0907	1.1261	3.9542	1.1890	1.8815	5.8175		2.5019
1.03–1.06	0.9630	1.2113	3.1659	1.0692	2.0258	7.2277		2.4130
>1.06	0.4576	1.0254	5.8086	0.6059	1.9338	6.0313		2.4197
Overall	0.8267	1.1233	3.9136	0.9964	1.5837	3.4962		1.9587

Examining the results in the top half of the table, we learn that for LIFFE options, regardless of whether the underlying asset pays a dividend, Heston's model gives smaller overall errors than either the Black-Scholes or MT models. We also see that for SPX options from either 2009 or 2010, the MT model's overall errors are the smallest. We see a clear difference in model performance between index and individual equity options, justifying the inclusion of both data smets in this work.

Looking at the results in the top half of the table in more detail, we see:

- For the entire LIFFE data set that includes dividend-paying underlying assets (LIFFE118), Heston's model does not outperform the MT model in all moneyness-maturity bins. However, if we examine only the LIFFE data for options on non-dividend-paying underlying assets (LIFFE25), Heston's model has substantially smaller out-of-sample pricing errors than the MT model in all bins.

- For the LIFFE118 data set, the overall out-of-sample pricing errors for Heston's model are 22% and 7% less than for the Black-Scholes and MT models, respectively. For the LIFFE25 data set, the overall out-of-sample pricing errors for Heston's model are 49% and 38% less than for the Black-Scholes and MT models, respectively. Finally, Heston's model does not possess the best overall errors for SPX options, and in this case the underlying asset does pay dividends.
- Putting these results together, we see that the performance of Heston's model differs for options on dividend-paying versus non-dividend-paying underlying assets.
- For the two LIFFE data sets, the MT model has a smaller out-of-sample pricing error than the Black-Scholes model in*every* moneyness-maturity bin. For the SPX data sets, the MT model has a smaller out-of-sample pricing error than the Black-Scholes model in most moneyness-maturity bins. This confirms and substantially extends prior results on the superiority of the MT model to the Black-Scholes model [1,2].

In the bottom half of Table 7, we present out-of-sample pricing errors for the practitioner's Black-Scholes method, the robust MT method, and the practitioner's MT method. The main finding here is that the practitioner's MT method's performance is very close to that of the practitioner's Black-Scholes method. In many bins, the practitioner's MT method gives errors that are smaller by a few cents, but never by a large margin. At the same time, both practitioner's methods clearly outperform the robust MT method.

Comparing these out-of-sample pricing results against those in the top half of Table 7, we reach the following conclusions:

- Across all four data sets, the practitioner's MT scheme clearly outperforms both the robust MT method as well as the overconstrained L2L2 method for the MT model. This is true not only for overall errors, but is also true for nearly every moneyness-maturity category. While strike- and maturity-dependent volatilities are inconsistent with the assumptions in the MT stochastic model, they do lead to the smallest out-of-sample errors. For those whose sole objective is to minimize out-of-sample pricing errors, the practitioner's MT method is the best of the three MT methods.

Table 7. For each of six models and four data sets (given in boldface), we provide out-of-sample pricing errors binned by moneyness and maturity. Rows and columns separate options by moneyness (m=S0/K) and maturity (time to expiration in days), respectively. Errors for LIFFE and SPX options are in Euros (€) and USD ($), respectively. The results are discussed in detail in Section 8.2.

	Black-Scholes				**Markov Tree**				**Heston**			
	<60	**60–180**	**>180**	**Overall**	**<60**	**60–180**	**>180**	**Overall**	**<60**	**60–180**	**>180**	**Overall**
						LIFFE118						
<0.94	0.0961	0.1827	0.3340	0.2216	0.0887	0.1550	0.2711	0.1849	0.0777	0.1393	0.2991	0.1890
0.94–0.97	0.1532	0.2286	0.3255	0.2361	0.1468	0.2194	0.2881	0.2175	0.1195	0.1523	0.2900	0.1911
0.97–1	0.1791	0.2408	0.3158	0.2447	0.1750	0.2388	0.2924	0.2340	0.1352	0.1585	0.3074	0.2044
1–1.03	0.1928	0.2623	0.3150	0.2540	0.1882	0.2590	0.3023	0.2468	0.1318	0.1638	0.3285	0.2105
1.03–1.06	0.1929	0.2857	0.3162	0.2610	0.1802	0.2716	0.3065	0.2490	0.1182	0.1712	0.3189	0.2045
>1.06	0.1439	0.3396	0.4044	0.3042	0.1245	0.2297	0.3373	0.2403	0.0667	0.1747	0.3753	0.2222
Overall	0.1443	0.2723	0.3619	0.2661	0.1319	0.2127	0.3079	0.2242	0.0892	0.1611	0.3367	0.2076
						LIFFE25						
<0.94	0.0705	0.1379	0.1934	0.1402	0.0656	0.1146	0.1555	0.1165	0.0471	0.0727	0.0919	0.0728
0.94–0.97	0.0962	0.1407	0.1661	0.1332	0.0929	0.1304	0.1470	0.1224	0.0643	0.0756	0.0933	0.0777
0.97–1	0.1055	0.1377	0.1682	0.1370	0.1028	0.1330	0.1561	0.1303	0.0680	0.0735	0.1010	0.0816
1–1.03	0.1059	0.1420	0.1630	0.1361	0.1024	0.1368	0.1547	0.1304	0.0647	0.0732	0.1002	0.0798
1.03–1.06	0.1067	0.1540	0.1644	0.1403	0.0990	0.1441	0.1572	0.1322	0.0588	0.0733	0.1001	0.0778
>1.06	0.0845	0.1867	0.2053	0.1622	0.0713	0.1199	0.1652	0.1228	0.0356	0.0712	0.1110	0.0760
Overall	0.0866	0.1620	0.1921	0.1493	0.0782	0.1223	0.1596	0.1226	0.0472	0.0723	0.1024	0.0759

						SPX09						
<0.94	1.0283	1.8973	3.3559	1.3307	0.9923	1.9481	3.0750	1.2952	1.1567	2.4181	2.6902	1.4810
0.94–0.97	1.9542	2.3008	4.2466	2.1403	1.7276	2.3129	3.8562	1.9460	2.0058	2.5075	2.6868	2.1301
0.97–1	1.7818	2.1510	4.4395	2.0075	1.6928	2.1234	4.1764	1.9238	1.9189	2.2149	3.0285	2.0465
1–1.03	1.7268	1.8951	4.0414	1.8345	1.6345	1.7578	4.1126	1.7383	1.6763	1.9303	3.3505	1.7809
1.03–1.06	1.6237	2.1759	3.6915	1.7820	1.5654	1.8046	4.1257	1.6818	1.4313	1.8417	3.0522	1.5514
>1.06	1.1800	1.6689	3.9170	1.2961	1.2352	1.5369	3.3974	1.3095	1.1188	1.4671	4.6104	1.2100
Overall	1.4704	1.9532	3.8964	1.6536	1.4174	1.8569	3.7149	1.5874	1.4694	1.9887	2.9690	1.6264
						SPX10						
<0.94	2.5247	4.6013		3.1887	2.0081	3.9160		2.6181	1.1167	1.6214		1.2780
0.94–0.97	2.5631	3.5518		2.6935	2.1082	3.3628		2.2737	2.5404	3.0072		2.6020
0.97–1	2.2363	4.2364		2.5801	2.0710	3.9689		2.3973	3.9287	2.7678		3.7291
1–1.03	2.3524	5.6459		2.8715	2.0468	4.6468		2.4566	4.2024	2.5562		3.9429
1.03–1.06	2.9328	9.0714		3.3898	1.9820	6.4140		2.3119	3.7052	2.7069		3.6309
>1.06	2.5203	9.7335		3.3761	2.1347	5.8657		2.5774	2.4940	2.9696		2.5504
Overall	2.4963	5.1198		3.0107	2.0573	4.1872		2.4750	2.7332	2.1798		2.6247

	Practitioner's Black-Scholes				Robust Markov Tree				Practitioner's Markov Tree			
	<60	60–180	>180	Overall	<60	60–180	>180	Overall	<60	60–180	>180	Overall
LIFFE118												
<0.94	0.0242	0.0335	0.0330	0.0306	0.0769	0.1316	0.2826	0.1796	0.0242	0.0335	0.0330	0.0306
0.94–0.97	0.0323	0.0418	0.0400	0.0375	0.1286	0.1838	0.2904	0.2026	0.0322	0.0417	0.0395	0.0373
0.97–1	0.0358	0.0450	0.0431	0.0408	0.1540	0.2011	0.2963	0.2180	0.0355	0.0448	0.0422	0.0403
1–1.03	0.0351	0.0456	0.0457	0.0415	0.1662	0.2214	0.3071	0.2306	0.0347	0.0452	0.0442	0.0407
1.03–1.06	0.0318	0.0455	0.0463	0.0405	0.1600	0.2359	0.3107	0.2338	0.0317	0.0449	0.0449	0.0398
>1.06	0.0208	0.0571	0.0895	0.0587	0.1110	0.2067	0.3541	0.2365	0.0239	0.0437	0.0694	0.0478
Overall	0.0260	0.0471	0.0615	0.0459	0.1165	0.1860	0.3198	0.2166	0.0272	0.0409	0.0521	0.0409
LIFFE25												
<0.94	0.0187	0.0265	0.0219	0.0223	0.0574	0.0966	0.1647	0.1126	0.0186	0.0266	0.0220	0.0223
0.94–0.97	0.0225	0.0299	0.0247	0.0253	0.0810	0.1083	0.1491	0.1128	0.0225	0.0299	0.0247	0.0252
0.97–1	0.0244	0.0311	0.0250	0.0264	0.0896	0.1089	0.1637	0.1219	0.0244	0.0311	0.0251	0.0264
1–1.03	0.0237	0.0320	0.0252	0.0264	0.0894	0.1143	0.1594	0.1214	0.0236	0.0320	0.0252	0.0264
1.03–1.06	0.0220	0.0315	0.0251	0.0256	0.0864	0.1208	0.1612	0.1229	0.0220	0.0314	0.0251	0.0256
>1.06	0.0150	0.0281	0.0242	0.0224	0.0625	0.1035	0.1724	0.1182	0.0161	0.0264	0.0238	0.0221
Overall	0.0184	0.0284	0.0238	0.0233	0.0684	0.1038	0.1667	0.1170	0.0189	0.0276	0.0236	0.0232
SPX09												
<0.94	0.7297	1.4533	1.8435	0.9071	0.9699	1.9284	3.5258	1.3027	0.7289	1.4535	1.8470	0.9067
0.94–0.97	1.1546	1.5697	1.7907	1.2499	1.7557	2.3093	4.4246	1.9983	1.1513	1.5673	1.7941	1.2470
0.97–1	1.1518	1.3848	1.3618	1.2105	1.7074	2.1721	4.8128	1.9790	1.1478	1.3820	1.3632	1.2071
1–1.03	1.1573	1.2459	1.6488	1.1875	1.6111	1.7243	4.8980	1.7389	1.1457	1.2469	1.6519	1.1784
1.03–1.06	1.1027	1.3394	1.3901	1.1400	1.5343	1.9275	4.8187	1.7028	1.0785	1.3287	1.3939	1.1183
>1.06	0.9762	0.9922	2.4322	0.9880	1.1814	1.6359	3.7108	1.2851	0.9005	0.9412	2.4268	0.9175
Overall	1.0227	1.3044	1.6695	1.0946	1.3978	1.8976	4.3047	1.6019	1.0014	1.2913	1.6722	1.0756

- On each of the four data sets, the practitioner's MT method makes significantly smaller out-of-sample pricing errors than Heston's model. This is true not only for overall errors, but also in each individual moneyness-maturity category.
- Between the two overconstrained methods for fitting the MT model, the robust method has smaller overall errors on the LIFFE data sets and on the 2010 SPX data set, while the L2L2 method has smaller overall errors on the 2009 SPX data set. For LIFFE options, if we focus our attention on short- and medium-term options, the robust method performs noticeably better than the L2L2 method; the L2L2 method is the superior method for long-term options. These statements do not carry over to the SPX data set, indicating again the difference between tests for individual equity options and index options.

9. CONCLUSIONS

Because the MT model consistently yields the least out-of-sample hedging errors, in overall categories and in almost all moneyness-maturity bins, the MT model should be considered a viable alternative to established models such as Heston's model and the Black-Scholes model.

The MT model achieves its superior hedging performance with two fewer regression coefficients than Heston's model. While the in-sample pricing errors decrease as a function of model complexity, the out-of-sample hedging errors are minimized by a model with three parameters (MT) rather than five (Heston), implying that Heston's model overfits the data.

We found that Heston's model prices options more accurately when the option's underlying asset pays no dividend. This contrasts with the MT model, which hedges equally well for options on dividend-paying and non-dividend-paying assets.

Suppose we take the point of view that Heston's model is not the best stochastic volatility model. We can still conclude that accounting for short-term memory in the underlying asset process—even in the simple, discrete-time setting of the MT model—yields significant reductions in out-of-sample hedging errors. This conclusion can be applied to any model. For instance, one could build short-term memory of the underlying asset into a sophisticated stochastic volatility model, thereby producing a model that may be superior to the MT model considered here.

Among all the methods we considered, the practitioner's MT method makes the smallest out-of-sample pricing and hedging errors. This procedures allows the MT model parameters to depend on an option's strike and maturity, leading to an increase in the number of effective model parameters, and escaping the assumptions/constraints of the underlying asset process (7). Therefore, the practitioner's MT method is the method of choice for those who seek to minimize errors leaving other considerations aside.

This procedure of applying the previous day's strike/maturity-dependent implied parameters to pricing or hedging tasks on the current day could, in principle, be carried out for any option pricing model, including Heston's stochastic volatility model. However, our study shows that while this procedure leads to substantial improvements in out-of-sample pricing for both the Black-Scholes and MT models, it does not improve both models' hedging performance equally. The practitioner's MT model makes much smaller out-of-sample hedging errors than the practitioner's Black-Scholes model.

Four of the methods we considered do not allow model parameters to depend on an option's strike and maturity. Among these methods, the robust MT method makes the smallest out-of-sample hedging errors, at practically no additional computational cost compared to the overconstrained L2L2 MT method. Compared to the practitioner's MT method, the robust MT method sacrifices performance for parsimony. The particular loss function (15) we considered is one of many robust loss functions that can be used; our results showing that the regression residuals follow a distribution with heavier-than-Gaussian tails indicate that some robust regression technique for fitting the MT model may yield a much larger improvement over the L2L2 method.

One of the advantages of considering a data set that spans several years is that one can assess long-term trends in model performance. Over the period of time we considered, Heston's model produces hedges that, on a daily basis, are less consistent and hence riskier than the hedges produced by the MT model. This conclusion, which stems from fine-grained results displayed in Figure 1 and discussed in Section 6.1, are important from a risk management perspective. We prefer a model that is consistently wrong by a small amount to a model that is wildly inconsistent, even if it is occasionally more accurate. An overly narrow focus on errors averaged across time obfuscates this point.

Option prices themselves are themselves uncertain. We used bootstrapping to quantify how the uncertainty in option prices is propagated to the regression coefficients. What we found is that the MT model achieves better hedges than Heston's model, even with noisy regression coefficients that are estimated from option prices with generalized hyperbolic noise. This is a rigorous test of the MT model's performance, indicating again its consistency.

There are other insights that we take away from our analysis of the data. Regarding the general methods by which option pricing models are tested, we found that pricing errors by themselves are not indicative of the out-of-sample hedging errors committed by any of the three models: Black-Scholes, Heston, or MT. In a similar vein, the results obtained by analyzing S&P 500 index options do not by themselves indicate what happens when we analyze LIFFE individual equity options. Overall, the large volume of data we have analyzed has led to more conclusive insights into the MT model than would have been possible with smaller data sets. Given the results obtained here, we welcome further large-scale testing of the MT model.

ACKNOWLEDGMENTS

The authors gratefully acknowledge support for this research from UC Merced's Graduate Research Council, Graduate Division, and Applied Mathematics Unit.

AUTHOR CONTRIBUTIONS

Both authors (H. S. Bhat and N. Kumar) contributed equally to all aspects of this paper, including (i) conception and design of the study, (ii) derivation of analytical formulas and writing of computer codes; (iii) statistical analysis of data; and (iv) writing of the paper.

CONFLICTS OF INTEREST

The authors declare no conflict of interest.

REFERENCES

1. Bhat, H.S.; Kumar, N. Markov tree options pricing. In Proceedings of the Fourth SIAM Conference on Mathematics for Industry (MI '09), San Francisco, CA, USA, 9–10 October 2009; pp. 162–173.

2. Bhat, H.S.; Kumar, N. Option Pricing Under a Normal Mixture Distribution Derived from the Markov Tree Model.*Eur. J. Oper. Res.* **2012**, *223*, 762–774.

3. Black, F.; Scholes, M.S. The Pricing of Options and Corporate Liabilities.*J. Polit. Econ.* **1973**, *81*, 637–654.

4. Heston, S.L. A closed-form solution for options with stochastic volatility with applications to bond and currency options. *Rev. Financ. Stud.* **1993**, *6*, 327–343.

5. Kaeck, A. Hedging Surprises, Jumps, and Model Misspecification: A Risk Management Perspective on Hedging S&P 500 Options. *Rev. Financ.* **2013**, *17*, 1535–1569.

6. Hobson, D.; Rogers, L.C.G. Complete models with stochastic volatility. *Math. Financ.* **1998**, *8*, 27–48.

7. Chang, M.H.; Youree, R.K. Infinite-dimensional Black-Scholes equation with hereditary structure. *Appl. Math. Optim.***2007**, *56*, 395–424.

8. Arriojas, M.; Hu, Y.; Mohammed, S.E.; Pap, G. A delayed Black and Scholes formula. *Stoch. Anal. Appl.* **2007**, *25*, 471–492

9. Wu, M.; Huang, N.; Zhao, C. European option pricing with time delay. In Proceedings of the 2008 Chinese Control Conference (CCC), Kunming, China, 16–18 July 2008; IEEE: Piscataway, NJ, USA, 2008; pp. 589–593.

10. McWilliams, N.; Sabanis, S. Arithmetic Asian options under stochastic delay models. *Appl. Math. Financ.* **2011**, *18*, 423–446.

11. Appleby, J.A.D.; Riedle, M.; Swords, C. Bubbles and crashes in a Black-Scholes model with delay. *Financ. Stoch.* **2013**,*17*, 1–30.

12. Appleby, J.A.D.; Daniels, J.A. Long Memory and Financial Market Bubble Dynamics in Affine Stochastic Differential Equations with Average Functionals. 9 October 2013. arXiv:1310.2339 [math.PR].

13. Appleby, J.A.D.; Krol, K.; Daniels, J.A. A Black-Scholes model with long memory. 24 February 2012. arXiv:1202.5574 [q-fin.PR].

14. Kim, I.J.; Kim, S. Empirical comparison of alternative stochastic volatility option pricing models: Evidence from Korean KOSPI 200 index options market. *Pac.-Basin Financ. J.* **2004**, *12*, 117–142.

15. Aït-Sahalia, Y.; Kimmel, R. Maximum likelihood estimation of stochastic volatility models. *J. Financ. Econ.* **2007**, *83*, 413–452.

16. Bakshi, G.; Kapadia, N. Delta-hedged gains and the negative market volatility risk premium. *Rev. Financ. Stud.* **2003**,*16*, 527–566.

17. Christoffersen, P.; Heston, S.; Jacobs, K. The shape and term structure of the index option smile: Why multifactor stochastic volatility models work so well. *Manag. Sci.* **2009**, *55*, 1914–1932.

18. AitSahlia, F.; Goswami, M.; Guha, S. American option pricing under stochastic volatility: An empirical evaluation.*Comput. Manag. Sci.* **2010**, *7*, 189–206.

19. Nguyen, A.; Solvang, J.B. Empirical Comparison of Alternative Option Pricing Models. Master's Thesis, Copenhagen Business School, Copenhagen, Denmark, 2013.

20. Bates, D. Empirical option pricing: A retrospection. *J. Econom.* **2003**, *116*, 387–404.

21. Garcia, R.; Ghysels, E.; Renault, E. The Econometrics of Option Pricing. Available online: http://papers.ssrn.com/sol3/papers.cfm?abstract_id=463860 (accessed on 11 March 2013).

22. Broadie, M.; Detemple, J.B. Option Pricing: Valuation Models and Applications. *Manag. Sci.* **2004**, *50*, 1145–1177.

23. Alexander, C.; Kaeck, A.; Nogueira, L.M. Model risk adjusted hedge ratios. *Jo. Futures Mark.* **2009**, *29*, 1021–1049.

24. Brigo, D.; Mercurio, F. Lognormal-Mixture Dynamics and Calibration to Market Volatility Smiles. *Int. J. Theor. Appl. Financ.* **2002**, *5*, 427–446.

25. Lo, C.C.; Skindilias, K. An improved Markov chain approximation methodology: Derivatives pricing and model calibration. *Int. J. Theor. Appl. Financ.* **2014**, *17*.

26. Nandi, S. *Pricing and Hedging Index Options Under Stochastic Volatility: An Empirical Examination*; Working Paper 96-9; Federal Reserve Bank of Atlanta: Atlanta, GA, USA, 1996.

27. Bakshi, G.; Cao, C.; Chen, Z. Empirical Performance of Alternative Options Pricing Models. *J. Financ.* **1997**, *52*, 2003–2049.

28. Nandi, S. How important is the correlation between returns and volatility in a stochastic volatility model? Empirical evidence from pricing and hedging in the S&P 500 index options market. *J. Bank. Financ.* **1998**, *22*, 589–610.

29. Bakshi, G.; Cao, C.; Chen, Z. Pricing and hedging long-term options. *J. Econom.* **2000**, *94*, 277–318

30. An, Y.; Suo, W. An empirical comparison of option-pricing models in hedging exotic options. *Financ. Manag.* **2009**, *38*, 889–914.

31. Skindilias, K.; Lo, C.C. Local volatility calibration during turbulent periods. *Rev. Quant. Financ. Account.* **2015**, *44*, 425–444.

32. Rubinstein, M. Nonparametric tests of alternative option pricing models using all reported trades and quotes on the 30 most active CBOE option classes from August 23, 1976 through August 31, 1978. *J. Financ.* **1985**, *40*, 455–480.

33. Lamoureux, C.G.; Lastrapes, W.D. Forecasting stock-return variance: Toward an understanding of stochastic implied volatilities. *Rev. Financ. Stud.* **1993**, *6*, 293–326.

34. Hull, J.C.; White, A.D. The Pricing of Options on Assets with Stochastic Volatilities. *J. Financ.* **1987**, *42*, 281–300.

35. Bakshi, G.; Kapadia, N.; Madan, D. Stock Return Characteristics, Skew Laws, and the Differential Pricing of Individual Equity Options. *Rev. Financ. Stud.* **2003**, *16*, 101–143.

36. Friesen, G.C.; Zhang, Y.; Zorn, T.S. Heterogeneous beliefs and risk-neutral skewness. *J. Financ. Quant. Anal.* **2012**, *47*, 851–872.

37. Chou, R.K.; Chung, S.L.; Hsiao, Y.J.; Wang, Y.H. The impact of liquidity on option prices. *J. Futures Mark.* **2011**, *31*, 1116–1141.

38. Yan, S. Jump risk, stock returns, and slope of implied volatility smile. *J. Financ. Econ.* **2011**, *99*, 216–233.

39. Chang, C.C.; Chou, P.H.; Liao, T.H. Fitting and testing for the implied volatility curve using parametric models. *J. Futures Mark.* **2012**, *32*, 1171–1191.

40. Gilli, M.; Schumann, E. Calibrating the Heston Model with Differential Evolution. In *Applications of Evolutionary Computation*; Lecture Notes in Computer Science; Springer: Berlin, Germany; Heidelberg, Germany, 2010; Volume 6025, pp. 242–250.

41. R Core Team. *R: A Language and Environment for Statistical Computing*; R Foundation for Statistical Computing: Vienna, Austria, 2012. Available online: http://www.R-project.org/ (accessed on 11 March 2013).

42. Mikhailov, S.; Nögel, U. Heston's stochastic volatility model implementation, calibration and some extensions. *Wilmott Mag.* **2003**, *3*, 74–79.

43. Fiorentini, G.; Leon, A.; Rubio, G. Estimation and empirical performance of Heston's stochastic volatility model: The case of a thinly traded market. *J. Empir. Financ.* **2002**, *9*, 225–255.

44. West, G. Calibration of the SABR model in illiquid markets. *Appl. Math. Financ.* **2005**, *12*, 371–385.

45. Hartley, R.; Zisserman, A. *Multiple View Geometry in Computer Vision*, 2nd ed.; Cambridge University Press: Cambridge, UK, 2003.

46. Byrd, R.H.; Lu, P.; Nocedal, J.; Zhu, C. A limited memory algorithm for bound constrained optimization. *SIAM J. Sci. Comput.* **1995**, *16*, 1190–1208.

47. Ypma, J. *Nloptr: R Interface to NLopt*, 2014. Available online: http://cran.r-project.org/package=nloptr (accessed on 27 April 2015).

48. Johnson, S.G. *The NLopt Nonlinear-Optimization Package*, 2013. Available online: http://ab-initio.mit.edu/nlopt (accessed on 27 April 2015).

49. Conn, A.R.; Gould, N.I.M.; Toint, P.L. A globally convergent augmented Lagrangian algorithm for optimization with general constraints and simple bounds. *SIAM J. Numer. Anal.* **1991**, *28*, 545–572.

50. Birgin, E.G.; Martinez, J.M. Improving ultimate convergence of an augmented Lagrangian method. *Optim. Methods Softw.* **2008**, *23*, 177–195.

51. Zhao, B.; Hodges, S.D. Parametric modeling of implied smile functions: A generalized SVI model. *Rev. Deriv. Res.* **2013**, *16*, 53–77.

52. Dickinson, K. *Financial Markets Operations Management*; Wiley: Chichester, UK, 2015.

53. McNeil, A.J.; Frey, R.; Embrechts, P. *Quantitative Risk Management: Concepts, Techniques, and Tools*; Princeton Series in Finance; Princeton University Press: Princeton, NJ, USA, 2005.

54. Luethi, D.; Breymann, W. *Ghyp: A Package on Generalized Hyperbolic Distributions*, 2013. Available online: http://cran.r-project.org/package=ghyp (accessed on 27 April 2015).

4

CHAPTER

Quadratic Hedging of Basis Risk

Hardy Hulley [1] and Thomas A. McWalter [2,3]

[1]*University of Technology Sydney, Finance Discipline Group, P.O. Box 123, Broadway, NSW 2007, Australia*

[2]*Department of Actuarial Science and the African Collaboration for Quantitative & Risk Research, University of Cape Town, Rondebosch, 7701, South Africa*

[3]*Faculty of Economic and Financial Sciences, Department of Finance and Investment Management, University of Johannesburg, P.O. Box 524, Auckland Park, 2006, South Africa*

ABSTRACT

This paper examines a simple basis risk model based on correlated geometric Brownian motions. We apply quadratic criteria to minimize basis risk and hedge in an optimal manner. Initially, we derive the Föllmer–Schweizer decomposition for a European claim. This allows pricing and hedging under the minimal martingale measure, corresponding to the local risk-minimizing strategy. Furthermore, since the mean-variance tradeoff process is deterministic in our setup, the minimal martingale- and variance-optimal martingale measures coincide. Consequently, the mean-variance optimal strategy is easily constructed. Simple pricing and hedging formulae for put and call options are derived in terms of the Black–Scholes formula. Due to market incompleteness, these formulae depend on the drift parameters of the processes. By making a further equilibrium assumption, we derive an approximate hedging formula, which does not require knowledge of these parameters. The hedging strategies are tested using Monte Carlo experiments, and are compared with results achieved using a utility maximization approach.

Keywords: option hedging; incomplete markets; basis risk; local risk minimization; mean-variance hedging

1. INTRODUCTION

When a contingent claim is written on an asset or process that is not traded, it is natural to enquire about the effectiveness of hedging with a correlated security. In this situation the market is incomplete, and the risk that arises as a result of imperfect hedging is known as basis risk. Examples include weather derivatives, real options, options on illiquid stocks and options on large baskets of stocks.

Since not all the risk can be hedged, we are dealing with a typical incomplete market situation, in which the appetite for risk must be specified (usually in terms of a utility function). A number of authors have formulated the problem of hedging basis risk in terms of the utility maximization approach (see, e.g., Davis [1,2], Henderson [3], Henderson and Hobson [4], Monoyios [5,6] and Zariphopoulou [7]). By contrast, we shall consider the application of quadratic criteria. Our approach is similar to that of Schweizer [8] (see also Duffie and Richardson [9]), where the application was hedging futures with a correlated asset. For comprehensive reviews on the theory of quadratic hedging, the reader is directed to the works of Pham [10], Schweizer [11] and McWalter [12].

We now describe the organization of this paper. To start with, in Section 2 we provide a summary of the terminology and general theory used throughout. In particular, we discuss the two quadratic approaches of local risk minimization and mean-variance hedging. Key to the construction of hedging strategies is a decomposition of the contingent claim, known as the Föllmer–Schweizer (FS) decomposition. We also briefly describe the minimal martingale measure and the variance-optimal martingale measure.

A simple basis risk model comprising two correlated geometric Brownian motions is specified in Section 3. We assume that it is not possible to trade in the process the claim is written on, but that the second process is a security available for trade. Since the general theory is developed in terms of discounted securities, we specify the discounted dynamics of the two processes.

The FS decomposition of the claim is derived in Section 4. This is achieved by expressing the non-traded process in terms of the traded security and an orthogonal process. By using a drift-adjusted representation of the non-traded process, it is possible to construct the minimal martingale measure. The Feynman–Kac theorem can then be used to express the discounted claim price as the solution of a partial differential equation (PDE) boundary-value problem.

In Section 5 we present the hedging strategies for the two quadratic approaches. The FS decomposition makes it easy to specify the locally risk-minimizing strategy, with prices determined by taking expectations under the minimal martingale measure. Furthermore, since the mean-variance tradeoff process is deterministic under our model assumptions, the minimal martingale measure and the variance-optimal martingale measure coincide. The mean-variance optimal self-financing strategy is thus easily constructed as well.

Having obtained a PDE representation of the price of the claim and its hedge parameters, in discounted terms, Section 6does the same in non-discounted terms by employing a simple transformation of variables. Remarkably, the PDE that emerges, for both local risk minimization and mean-variance optimization, is the familiar Black–Scholes PDE, with a "dividend yield" parameter playing a risk-adjustment role. Consequently, both approaches yield classical closed-form derivative pricing formulae for European calls and puts (which is advantageous from a computational point of view). The hedge ratios for the two quadratic criteria are different, which reflects the different attitudes to risk they imply. These results are summarized in Proposition 3.

In Section 7 we briefly introduce the utility maximization approach to our problem that was proposed by Monoyios [5,6]. In the limiting case where risk is minimized, we observe that his hedging algorithm becomes the local risk-minimizing strategy.

A disadvantage of the quadratic hedging rules, as well as the approach of Monoyios, is that they rely explicitly on drift estimates for the asset price processes. In Section 8 we make an extra assumption, based on equilibrium under the capital asset pricing model (CAPM), which allows the derivation of a "naive" approximation of the local risk-minimizing strategy. The advantage of this naive strategy is that it does not require knowledge of the drift parameters.

In Section 9 we demonstrate the effectiveness of the quadratic hedging approaches numerically and compare the results with those obtained using the Monoyios scheme and the naive strategy. Finally, Section 10 concludes the paper.

2. GENERAL THEORY

This section briefly introduces the terminology and theory required for the remainder of the paper. We shall merely summarize the necessary results; the reader is directed to the literature, primarily the account of Schweizer [11], for further information and proofs.

We start by fixing a finite time-horizon T∈(0,∞) and a filtered probability space (Ω,F,F,P). All processes are defined on this space, exist over the time interval [0,T], and are adapted to the filtration F=(Ft) t∈[0,T], which in turn is assumed to obey the usual conditions. For the sake of simplicity, we take F0F0 to be trivial, and set F:=FT.

We consider a frictionless financial market, with a single risky security and a bank account, denoted by *S* and *B*respectively. The process *X* will represent the discounted risky security, *i.e.*, X:=S/B. For the moment, we leave the dynamics of the processes unspecified, except to say that the bank account is a predictable process with finite variation, and that *X* is a special semimartingale with canonical decomposition $X = X_0 + M + A$, where $M \in \mathcal{M}^2_{0,\mathrm{loc}}(\mathbb{P})$ and *A* is a process with finite variation. We say that *X* satisfies the *structure condition* if there exists a predictable process *α*, such that A=∫αd⟨M⟩ and the *mean-variance tradeoff process* $\widehat{K} := \int \alpha^2 \, d\langle M \rangle$ is a.s. finite. In addition, we also introduce a contingent claim *H*, which we take to be an FTFT-measurable square-integrable random variable.

Let Θ denote the family of predictable processes *ϕ*, such that the *gain process* $G(\phi) := \int \phi \, dX$ belongs to the space S2(P) of square-integrable semimartingales. A *hedging strategy* is a pair of processes (ξ,η), where ξ∈Θ and *η* is an adapted process, such that the *value process* V(ξ,η):=ξX+η is right continuous and square-integrable. The process *ξ*represents a holding in *X*, while *η* represents a holding in the bank account.

Since we are dealing with an incomplete market, the cost of a contingent claim is not unique, and is consequently preference-dependent. In order to quantify the risk of imperfect hedging, a *cost process* C(ξ,η):=V(ξ,η)−G(ξ) is introduced. In a complete market *C* is deterministic and equal to the preference-independent price of the claim—this follows directly from the martingale representation results of Harrison and Pliska [13,14]. In an incomplete market *C* is a stochastic process. Our aim is to price the claim by estimating CTCT at inception, and to minimize the risk (*i.e.*, the deviation of the hedge portfolio from the terminal payoff) by minimizing a suitable quadratic functional of the cost process. We briefly outline the two approaches of local risk minimization and mean-variance hedging.

2.1. Local Risk Minimization

With this approach we consider those strategies that replicate the contingent claim H at time T; *i.e.*, we insist on the condition

$$V_T(\xi, \eta) = H \quad \text{a.s.} \tag{1}$$

Since the market is incomplete, we need to relax the usual complete market constraint that the value process be self-financing. As it happens, the weaker notion of a *mean-self-financing* strategy—which corresponds to the situation where the cost process is a martingale—is appropriate in this context.

Local risk minimization is a variational concept. Intuitively, it entails the instantaneous minimization of the conditional variance of the increments of the cost function C under the measure PP. This is implemented with the introduction of a risk-quotient, which we do not consider here—for details we refer the reader to the original references (Schweizer [11,15,16]). Subject to certain technical conditions, it can be shown that finding the local risk-minimizing strategy is equivalent to finding a decomposition of the claim, known in the literature as the Föllmer–Schweizer decomposition. A claim H is said to admit a *Föllmer-Schweizer* (*FS*) *decomposition* if it can be expressed as

$$H = H_0 + \int_0^T \xi_s^H \, dX_s + L_T^H \quad \text{a.s.,} \tag{2}$$

where $H_0 \in \mathbb{R}$, $\xi^H \in \Theta$ and $L^H \in \mathcal{M}_0^2(\mathbb{P})$ is strongly orthogonal to M.

In order to provide a precise statement of the local risk minimization optimality result, we need to introduce the so-called minimal martingale measure. To simplify matters, we shall assume that X, and hence also M and A, are continuous. Now, define a process Zˆ Z^, by setting

$$\widehat{Z}_t := \mathscr{E}\left(-\int_0^{\cdot} \alpha_s \, dM_s\right)_t,$$

for all $t \in [0,T]$. It can be shown (see Schweizer [11]) $\widehat{Z} \in \mathcal{M}^2_{0,\mathrm{loc}}(\mathbb{P})$, and that Z^X and Z^L are PP-local martingales, for all L∈M20,loc(P) strongly orthogonal to *M*. Note that continuity of *M* and the assumption of a.s. finite K^ ensure that the Doléans exponential above is strictly positive. Now suppose, furthermore, that $\widehat{Z} \in \mathcal{M}^2(\mathbb{P})$, and define a probability measure P^~P by setting

$$\frac{d\widehat{\mathbb{P}}}{d\mathbb{P}} := \widehat{Z}_T \in L^2(\mathbb{P}).$$

Then Z^Z^ may be interpreted as the density process for P^P^, in the sense that $d\widehat{\mathbb{P}}/d\mathbb{P}|_{\mathscr{F}_t} = \widehat{Z}_t$, for all t∈[0,T].

The probability measure P^P^, defined above, is an equivalent local martingale measure (ELMM) for *X*, and is called the*minimal martingale measure*. It is minimal in the sense that, apart from transforming *X* into a local martingale, it preserves the remaining structure of the model—in particular it preserves the martingale property of all martingales strongly orthogonal to *M* (see Föllmer and Schweizer [17] for an amplification of this point). We are now able to state the optimality result.

Theorem 1. *Suppose that X is continuous, and therefore satisfies the structure condition (see Schweizer [11, p. 553] and [18, Theorem 1] for justification). Furthermore, suppose that the density process for the minimal martingale measure for X satisfies*

$$\widehat{Z} \in \mathcal{M}^2(\mathbb{P}). \tag{3}$$

If H admits an FS decomposition (2), then $(\widehat{\xi}, \widehat{\eta}) := (\xi^H, \widehat{V} - \xi^H X)$ *determines a (mean-self-financing) locally risk-minimizing strategy for H, where the intrinsic value process is defined by setting*

$$\widehat{V}_t := \mathbb{E}^{\widehat{\mathbb{P}}}\left[H \mid \mathscr{F}_t\right] = H_0 + G_t(\xi^H) + L_t^H,$$

for all t∈[0,T] *Here G is the gain process and* LHLH *may be interpreted as the unhedged risk. Furthermore, a sufficient condition for (2) and (3) is that the mean-variance tradeoff process,* K^K^, *is uniformly bounded.*

Proof. See Theorem 3.5 of Schweizer [11]. □

2.2. Mean-Variance Hedging

In contrast to local risk minimization, we now insist that the hedge portfolio be self-financing over the life of the option [0,T)[0,T). At its maturity, however, a profit or shortfall is realized, so that condition (1) is met. The mean-variance optimal strategy is characterized as that strategy for which the profit or loss at time *T* has the smallest variance. More precisely, the*mean-variance optimal strategy* is the self-financing strategy (ξ˜,η˜), with $(\widetilde{\xi}, \widetilde{\eta})$, with $\widetilde{\xi} = \xi^{(v)}$ and $V_0(\widetilde{\xi}, \widetilde{\eta}) = v,$ such that

$$\mathbb{E}\left[(C_T - v)^2\right] = \mathbb{E}\left[\left(H - v - G_T(\xi^{(v)})\right)^2\right]$$

is minimized over all $\left(v, \xi^{(v)}\right) \in \mathbb{R} \times \Theta.$. The initial value *v* is known as the *approximation price* of *H*.

Related to the problem of finding the mean-variance optimal strategy is the problem of finding the variance-optimal martingale measure for *X*. As before, we assume that *X* is continuous and consider the set P2e(X) of all measures Q~P where Q is an ELMM for *X*, with dQdP∈L2(P) (In the case where *X* is not continuous, a more general set of signed martingale measures must be considered—see Section 4 of Schweizer [11] for details.) Then a measure in P2e(X) is called*variance-optimal* if it minimizes

$$\mathrm{Var}\left[\frac{d\mathbb{Q}}{d\mathbb{P}}\right] = \mathbb{E}\left[\left(\frac{d\mathbb{Q}}{d\mathbb{P}} - 1\right)^2\right] = \mathbb{E}\left[\left(\frac{d\mathbb{Q}}{d\mathbb{P}}\right)^2\right] - 1,$$

over all $\mathbb{Q} \in \mathbf{P}_e^2(X).$.

In general, the minimal martingale measure and the variance optimal martingale measure are different, in which case significant effort is required to find the mean-variance optimal strategy (see, e.g., Heath *et al.* [19]). Under certain circumstances, however, the measures coincide—the following theorem provides such an instance, and the resultant form of the mean-variance optimal strategy.

Theorem 2. *Suppose X is continuous, and that* K^TK^T *is deterministic (thus ensuring that the FS decomposition (2) exists). Then the variance-optimal martingale measure and the minimal martingale measure coincide. Furthermore, the mean-variance optimal strategy for H is the self-financing strategy* $(\widetilde{\xi},\widetilde{\eta})$, *with* $\widetilde{\xi}=\xi^{(v)}$, where

$$v=\mathbb{E}^{\widehat{\mathbb{P}}}\left[H\right]=H_0$$

and

$$\xi_t^{(v)}=\xi_t^H+\alpha_t\left(\widehat{V}_t-v-G_t(\xi^{(v)})\right),$$

for all t∈[0,T]. *Here,* V^V^ *is the intrinsic value process (Theorem 1) and G is the gain process. It then easily follows from the self-financing property that*

$$\widetilde{\eta}_t=v+G_t\left(\xi^{(v)}\right)-\xi_t^{(v)}X_t,$$

for all t∈[0,T].

Proof. See Theorems 4.6 and 4.7 of Schweizer [11]. □

With the mathematical requisites established, we are now in a position to introduce the market assumptions and apply the theory to the problem of hedging basis risk.

3. MARKET ASSUMPTIONS

As in the previous section, we fix a finite time-horizon T∈(0,∞)T∈(0,∞) and a stochastic basis (Ω,F,F,P) which supports two orthogonal Brownian motions *W* and W⊥W⊥. All processes are defined on the above stochastic basis (in particular, they exist over the time interval [0,T], and are adapted to the filtration F=(Ft)t∈[0,T], which we take to be the augmentation of the filtration generated by *W* and W⊥. Therefore it follows that FF satisfies the usual conditions.

We specify a bank account process *B* as follows:

$$B_t := e^{rt},$$

for all t∈[0,T], where r>0 is a constant short rate. The process *S* represents a traded risky asset, while *U* is an observable correlated process on which an option is written. The option is European, with maturity *T* and payoff h(UT), for some Borel-measurable function h:R+→R+h: The objective is to hedge this claim using the traded asset *S*, in such a way that the basis risk is minimized.

Since the analysis is carried out using discounted assets, we introduce two new processes—*X* representing the discounted traded asset, and *Y* representing the discounted non-traded asset—by setting

$$X := \frac{S}{B} \quad \text{and} \quad Y := \frac{U}{B}.$$

Furthermore, we assume that the discounted assets are driven by the Brownian motions *W* and W⊥W⊥, as follows:

$$dX_t = (\mu_S - r)X_t\,dt + \sigma_S X_t\,dW_t, \qquad (4)$$

$$dY_t = (\mu_U - r)Y_t\,dt + \sigma_U Y_t(\rho\,dW_t + \sqrt{1-\rho^2}\,dW_t^{\perp}), \qquad (5)$$

for all $t \in [0,T]$, where the drifts μU,μSμU,μS, the volatilities σU,σS>0 and correlation −1≤ρ≤1 are constants. We wish to hedge the discounted European claim h¯(YT), where $\bar{h}(Y_T)$, where $\bar{h} : \mathbb{R}_+ \to \mathbb{R}_+$ is defined by

$$\bar{h}(x) := e^{-rT} h(e^{rT}x),$$

for all x∈R+x, using the discounted traded asset *X*.

For convenience, we define the Sharpe ratios for the traded asset and the non-traded process as follows:

$$\theta_S := \frac{\mu_S - r}{\sigma_S} \quad \text{and} \quad \theta_U := \frac{\mu_U - r}{\sigma_U}.$$

In the case where the assets are perfectly correlated (*i.e.*, $\rho=1\rho=1$), it is well known (see, e.g., Davis [1]) that the absence of arbitrage implies that their Sharpe ratios should be equal (*i.e.*, $\theta U=\theta S$). Under this condition, the (non-discounted) price of a European call or put on the non-traded asset is given by $BS(t,Ut,0,\sigma U)$, where

$$\mathsf{BS}(t,s,q,\sigma) := \delta\left(se^{-q(T-t)}N(\delta d_1) - Ke^{-r(T-t)}N(\delta d_2)\right)$$

is the standard Black–Scholes formula, with

$$d_1 := \frac{\ln(s/K) + (r - q + \sigma^2/2)(T-t)}{\sigma\sqrt{T-t}} \quad \text{and} \quad d_2 := d_1 - \sqrt{T-t}.$$

Here, $\delta=1\delta=1$ for a call and $\delta=-1$ for a put, while $K>0$ is the strike price, and $q\in Rq\in R$ is a dividend yield parameter. Usually the dividend yield applies to the stock on which the option is priced. Note that we have not modeled a dividend yield in our basis risk model. We will, however, require this more general Black–Scholes formula later. Hedging is then achieved by holding

$$\frac{\sigma_U U_t}{\sigma_S S_t}\Delta_{\mathsf{BS}}(t, U_t, 0, \sigma_U) \tag{6}$$

units of the traded asset *S* at each time $t\in[0,T]$, where

$$\Delta_{\mathsf{BS}}(t,s,q,\sigma) := \delta e^{-q(t-T)}N(\delta d_1)$$

is the usual Black–Scholes delta. It will be shown that the quadratic hedging approaches are consistent with this limiting regime.

4. THE FÖLLMER–SCHWEIZER DECOMPOSITION

We now derive the FS decomposition for the basis risk model presented in the previous section. To start with, note that Xsatisfies the structure condition, since its canonical decomposition takes the form

$$X_t = X_0 + M_t + \int_0^t \alpha_s \, d\langle M \rangle_s, \tag{7}$$

with

$$M_t := \int_0^t \sigma_S X_s \, dW_s \quad \text{and} \quad \alpha_t := \frac{\mu_S - r}{\sigma_S^2 X_t}, \tag{8}$$

for all t∈[0,T]t∈[0,T]. We seek a decomposition of the discounted claim of the form

$$\bar{h}(Y_T) =: H = H_0 + \int_0^T \xi_s^H \, dX_s + L_T^H \quad \text{a.s.}, \tag{9}$$

where $H_0 \in \mathbb{R}$, $\xi^H \in \Theta$ and $L^H \in \mathcal{M}_0^2(\mathbb{P})$ is strongly orthogonal to M.

Now, by rearranging (4), we get

$$dW_t = \frac{dX_t}{\sigma_S X_t} - \theta_S \, dt,$$

for all t∈[0,T]. Substituting this into (5) yields

$$\frac{dY_t}{Y_t} = (\mu_U - r - \rho \sigma_U \theta_S) \, dt + \sigma_U \left(\frac{\rho}{\sigma_S X_t} \, dX_t + \sqrt{1 - \rho^2} \, dW_t^{\perp} \right), \tag{10}$$

for all t∈[0,T]. We now specify the drift-adjusted process Y˜ ˜as the unique strong solution of the backward stochastic differential equation

$$d\widetilde{Y}_t = \widetilde{Y}_t \left(\frac{dY_t}{Y_t} + \kappa\, dt \right),$$

for all t∈[0,T], where

$$\kappa := \sigma_U \left(\rho\theta_S - \theta_U \right), \qquad (11)$$

and Y˜T=YT is the terminal condition (for existence and uniqueness see §2 of El Karoui *et al.* [20]). A simple calculation shows that

$$\widetilde{Y}_t = e^{-\kappa(T-t)} Y_t,$$

which, when substituted into (10), yields

$$d\widetilde{Y}_t = \sigma_U \widetilde{Y}_t \left(\frac{\rho}{\sigma_S X_t}\, dX_t + \sqrt{1-\rho^2}\, dW_t^{\perp} \right), \qquad (12)$$

for all t∈[0,T].

We now construct the minimal martingale measure for *X*. By the definition of the minimal martingale measure and the canonical decomposition (7), the density process for P^P^ is given by

$$\widehat{Z}_t = \mathscr{E}\left(-\int_0^{\cdot} \alpha_s\, dM_s \right)_t = \mathscr{E}(-\theta_S W)_t,$$

for all t∈[0,T]. Since *X* is a martingale under P^ we can define a new process W^W^ as

$$d\widehat{W}_t = dW_t + \theta_S\,dt,$$

for all t∈[0,T] Since W^Z^ is a martingale and $\langle \widehat{W} \rangle_t^{\widehat{\mathbb{P}}} = [\widehat{W}]_t = t$, for all $t \in [0,T]$, for Lévy's characterization of Brownian motion (see, e.g., Shreve [21], Thm. 4.6.4, p. 168) informs us that W^W^ is a Brownian motion under P^P^. Rewriting (4) and (12) in terms of W^W^ gives

$$dX_t = \sigma_S X_t\,d\widehat{W}_t$$

and

$$d\widetilde{Y}_t = \sigma_U \widetilde{Y}_t\big(\rho\,d\widehat{W}_t + \sqrt{1-\rho^2}\,dW_t^{\perp}\big), \tag{13}$$

for all t∈[0,T]. Note that W⊥W⊥ is strongly orthogonal to *M*, which means that its martingale property is preserved under the minimal martingale measure. Since ⟨U⟩P^t=[U]t=t , we see that *U* is also a Brownian motion under P^P^, again by Lévy's characterization of Brownian motion. This in turn means that the expression in brackets in (13) is a Brownian motion under P^P^.

We now use the Feynman–Kac theorem (see, e.g., Shreve [21], Thm. 6.4.1, p. 268) to infer a PDE representation for the claim. Define $F : [0,T] \times (0,\infty) \to \mathbb{R}_+$, by setting

$$F(t,x) := \mathbb{E}^{\widehat{\mathbb{P}}}\left[\bar{h}(\widetilde{Y}_T)\,\middle|\,\widetilde{Y}_t = x\right],$$

for all $(t,x) \in [0,T] \times (0,\infty)$. . Obviously we then have

$$F(T,x) = \bar{h}(x), \tag{14}$$

for all x∈(0,∞). According to the Feynman–Kac theorem, *F* satisfies the following PDE:

$$\frac{\partial F}{\partial t}(t,x) + \frac{1}{2}\sigma_U^2 x^2 \frac{\partial^2 F}{\partial x^2}(t,x) = 0, \qquad (15)$$

for all $(t,x) \in [0,T] \times (0,\infty)$, with terminal condition (14). Applying Itô's formula to the process $(F(t,\widetilde{Y}_t))_{t\in[0,T]}$ yields

$$\begin{aligned}\bar{h}(Y_T) = \bar{h}(\widetilde{Y}_T) = F(T,\widetilde{Y}_T) = F(0,\widetilde{Y}_0) &+ \int_0^T \frac{\partial F}{\partial x}(t,\widetilde{Y}_s)\,d\widetilde{Y}_s \\ &+ \int_0^T \left(\frac{\partial F}{\partial t}(s,\widetilde{Y}_s) + \frac{1}{2}\sigma_U^2 \widetilde{Y}_s^2 \frac{\partial^2 F}{\partial x^2}(s,\widetilde{Y}_t)\right) ds.\end{aligned}$$

Substituting (12) and (15) into this expression gives

$$\begin{aligned}\bar{h}(Y_T) = F(0,\widetilde{Y}_0) &+ \int_0^T \rho \frac{\sigma_U \widetilde{Y}_s}{\sigma_S X_s}\frac{\partial F}{\partial x}(s,\widetilde{Y}_s)\,dX_s \\ &+ \int_0^T \sigma_U \widetilde{Y}_s \sqrt{1-\rho^2}\frac{\partial F}{\partial x}(s,\widetilde{Y}_s)\,dW_s^\perp.\end{aligned}$$

This is the FS decomposition we have been looking for. Comparing terms with (9), we obtain

$$\begin{aligned} H_0 &= F(0,\widetilde{Y}_0) = \mathbb{E}^{\widehat{\mathbb{P}}}\left[\bar{h}(\widetilde{Y}_T)\right], \\ \xi_t^H &= \rho \frac{\sigma_U \widetilde{Y}_t}{\sigma_S X_t}\frac{\partial F}{\partial x}(t,\widetilde{Y}_t) \quad \text{and} \\ L_t^H &= \int_0^t \sigma_U \widetilde{Y}_s \sqrt{1-\rho^2}\,\frac{\partial F}{\partial x}(s,\widetilde{Y}_s)\,dW_s^\perp, \end{aligned} \qquad (16)$$

for all t∈[0,T]

.

5. HEDGING STRATEGIES

Now that we have the FS decomposition, we can easily determine the locally risk-minimizing strategy. By Theorem 1 it is the mean-self-financing strategy (ξˆ,ηˆ)(ξ^,η^) determined by

$$(\widehat{\xi}_t, \widehat{\eta}_t) := \left(\xi^H_t, \widehat{V}_t - \xi^H_t X_t\right),$$

where ξHξH is given by (16) and

$$\widehat{V}_t := \mathbb{E}^{\widehat{\mathbb{P}}}\left[\bar{h}(\widetilde{Y}_T) \,\middle|\, \mathscr{F}_t\right] = F(t, \widetilde{Y}_t),$$

for all t∈[0,T]t∈[0,T]. (Here ξˆξ^ specifies the holding in the discounted traded asset *X*, and ηˆη^ is the holding in the bank account.)

It is also an easy matter to find the mean-variance optimal strategy. Since *X* satisfies the structure condition, we can use (8) to obtain the mean-variance tradeoff process KˆK^ as follows:

$$\widehat{K}_t = \int_0^t \alpha_s^2 \, d\langle M\rangle_s = \int_0^t \left(\frac{\mu_S - r}{\sigma_S}\right)^2 ds = \theta_S^2 t,$$

for all t∈[0,T]t∈[0,T]. This is clearly bounded on [0,T][0,T], and thus by Theorem 2, we can express the self-financing mean-variance optimal strategy (ξ˜,η˜)(ξ˜,η˜) as follows:

$$(\widetilde{\xi}_t, \widetilde{\eta}_t) := \left(\xi^{(v)}_t, v + G_t(\xi^{(v)}) - \xi^{(v)}_t X_t\right),$$

where

$$v = \mathbb{E}^{\widehat{\mathbb{P}}}\left[\bar{h}(\widetilde{Y}_T)\right] = F(0, \widetilde{Y}_0) \quad \text{and} \quad \xi^{(v)}_t = \xi^H_t + \alpha_t\left(\widehat{V}_t - v - G_t(\xi^{(v)})\right),$$

for all t∈[0,T]. Here G(ξ(v)) is the gain from trading in the discounted asset *X*, using ξ(v)ξ(v).

6. EXPRESSIONS FOR PRICING AND HEDGING

In the previous two sections, we manipulated the discounted assets to obtain the FS decomposition of the discounted claim as well as the hedge portfolios (in terms of the discounted assets) for the two quadratic approaches. It is interesting to note that (15) looks similar to the discounted Black–Scholes PDE. By transforming variables, we now consider the situation without discounting.

Define the function V:[0,T]×(0,∞)→R+V:[0,T]×(0,∞)→R+, by setting

$$V(t,s) := e^{rt} F\left(t, e^{-rt} s e^{-\kappa(T-t)}\right),$$

for all (t,s)∈[0,T]×(0,∞), where κ is defined by (11). The PDE (15) may then be rewritten as

$$rV(t,s) = \frac{\partial V}{\partial t}(t,s) + (r-\kappa)s\frac{\partial V}{\partial s}(t,s) + \frac{1}{2}\sigma_U^2 s^2 \frac{\partial^2 V}{\partial s^2}(t,s),$$

for all (t,x)∈[0,T]×(0,∞) with the terminal condition, corresponding to (14), given by

$$V(T,s) = h(s),$$

for all s∈(0,∞)s∈(0,∞). When h(s)h(s) is the payoff of a put or a call, the solution of this PDE is given by the Black–Scholes option pricing formula for a stock with a continuous dividend yield κ

$$V(t,s) = \mathsf{BS}(t,s,\kappa,\sigma_U).$$

Noting that

$$\frac{\partial V}{\partial s}(t,s) = e^{-\kappa(T-t)} \frac{\partial F}{\partial x}\left(t, e^{-rt} s e^{-\kappa(T-t)}\right),$$

a simple substitution into (16) allows the computation of the hedge parameters for the optimal strategies. We now summarize the results with the following proposition:

Proposition 3. *Under the market assumptions of Section 3, the approximation price of the claim is given by*

$$v = V(0, U_0) = \mathsf{BS}(0, U_0, \kappa, \sigma_U).$$

The local risk-minimizing strategy is given by

$$\widehat{\xi}_t = \rho \frac{\sigma_U U_t}{\sigma_S S_t} \frac{\partial V}{\partial s}(t, U_t) = \rho \frac{\sigma_U U_t}{\sigma_S S_t} \Delta_{\mathsf{BS}}(t, U_t, \kappa, \sigma_U), \quad (17)$$

and the mean-variance optimal strategy is given, in feedback form, by

$$\begin{aligned} \widetilde{\xi}_t &= \widehat{\xi}_t + \frac{\mu_S - r}{\sigma_S^2 e^{-rt} S_t} \left(\widehat{V}_t - v - G_t(\widetilde{\xi}) \right) \\ &= \widehat{\xi}_t + \frac{\mu_S - r}{\sigma_S^2 e^{-rt} S_t} \left(\widehat{V}_t - v - \int_0^t \widetilde{\xi}_u \, d(e^{-ru} S_u) \right), \end{aligned} \quad (18)$$

with the intrinsic value given by

$$\widehat{V}_t = e^{-rt} V(t, U_t) = e^{-rt} \mathsf{BS}(t, U_t, \kappa, \sigma_U),$$

for all t∈[0,T].

Note that when *U* and *S* are perfectly correlated (*i.e.*, when ρ=1ρ=1), arbitrage considerations ensure that their respective market prices of risk must be equal. In this case, it follows from (11) that κ=0κ=0, which implies that (17) and (6) are the same. This demonstrates that the local risk minimization approach is consistent with the standard Black–Scholes hedge in the limiting case of a complete market.

It is possible to show that the mean-variance hedging strategy is also equivalent to the Black–Scholes strategy in the limiting case. Since we are dealing with a complete market, the cost process is constant and equal to the approximation price of the claim *v*. Therefore, the term in brackets in (18) is equal to zero, for all t∈[0,T], thereby showing that the mean-variance strategy is equal to the local risk-minimizing strategy (and, in turn, to the Black–Scholes hedge).

7. THE UTILITY FORMULATION OF MONOYIOS

In this section, we briefly outline the utility formulation of the basis risk problem presented by Monoyios [5,6] and highlight the connection between this approach and the local risk-minimizing approach. We do not present the full details of how the pricing and hedging rules are derived, but direct the reader to the original papers for details.

The basic problem of hedging basis risk may be expressed as a utility maximization problem as follows:

$$(\xi^*, \eta^*) = \underset{(\xi,\eta)}{\arg\max}\, \mathbb{E}\left[\mathcal{U}(V_T(\xi,\eta) + nh(U_T))\right], \quad (19)$$

where n is the number of options written. The utility function used by Monoyios is the exponential utility function

$$\mathcal{U}(x) = -\exp(-\gamma x),$$

for all $x\in(0,\infty)$), where $\gamma\in(0,1)\gamma\in(0,1)$ is the risk aversion parameter.

Unfortunately the problem (19) does not have a closed-form analytical solution, and Monoyios therefore proposes two approximate schemes to allow the computation of prices and hedge parameters. The first paper [5] uses a perturbation expansion, whereas the second paper [6] uses cumulant expansions in a similar manner. In both cases, prices and hedge parameters are effectively specified as power series expansions in terms of the following dimensionless parameter:

$$a := -\gamma(1-\rho^2)n.$$

(Note that in the first paper by Monoyios [5], the Taylor expansion is made in terms of $\epsilon=1-\rho 2$, but only even powers of ϵ appear in the expansion, and each term incorporates the relevant power of γ.) Monoyios assumes that $n=-1n=-1$, corresponding to a single short put position, thereby ensuring that $0<a<10<a<1$. In Proposition 3 of Monoyios [6], the indifference pricing formula is given by

$$p^{(n)}(t,u) = e^{-r(T-t)} \sum_{j=1}^{5} \frac{a^{j-1}}{j!} k_j(h(U_T)) + O(a^5), \quad (20)$$

where kjkj is the *j*th cumulant of the payoff under the minimal martingale measure, conditional on Ut=u. The optimal hedging strategy (see Proposition 2 and Corollary 1 of Monoyios [6]) is given by

$$\Delta_t^{(H)} = -n\rho \frac{\sigma_U U_t}{\sigma_S S_t} \frac{\partial p^{(n)}}{\partial u}(t, U_t). \quad (21)$$

The pricing and hedging rules in Monoyios [5] are equivalent to the above rules, with the sum evaluated only over the first four terms.

The first cumulant in (20) is given $k_1(h(U_T)) = \mathbb{E}^{\hat{\mathbb{P}}}\left[h(U_T) \,|\, U_t = u\right]$, which means that, when risk is minimized (*i.e.*,γ→0γ→0), the indifference price of the option is equal to the approximation price given in Proposition 3. Consequently, in this limit, the optimal hedging strategy (21) above coincides with the local risk-minimizing hedge ratio (17).

It should be noted that in order to provide a good level of hedging (*i.e.*, low variance of the distribution of profits and losses from hedging), the risk aversion parameter should be small (γ~0.01γ~0.01). As we shall see in Section 9, it is only feasible to carry out basis risk hedging when the correlation has an absolute value close to unity. This leads to a very small positive value for *a*. With this in mind, we expect to see similar numerical results for the utility formulation to those produced by the local risk-minimizing approach.

Due to a constraint in the utility maximization formulation (see Assumption 1 in [6]), the algorithms proposed by Monoyios cannot be used directly for pricing and hedging a call option. To overcome this shortcoming, Henderson and Hobson [4] suggest modeling the call using a static hedge consisting of put options. By contrast, the quadratic formulations in Proposition 3 are applicable to both puts and calls, without modification.

8. A NAIVE STRATEGY

While the hedging strategies derived in Section 6 have the advantage of yielding simple expressions based on the Black–Scholes equation, they implicitly (through the dividend yield parameter κ) rely on being able to estimate the drift coefficients of the asset price processes. As is well known this is very difficult to do. For an amplification of this point, see Rogers [22], where the error in the classical Merton model, as a result of parameter uncertainty, is compared with the error due to discrete rebalancing.

Monoyios [6,23] proposes a filtering approach to estimate the drift coefficients from observations of the asset prices and finds expressions that could in principle be used as the starting point for a scheme similar to that outlined in the previous section. Here we pursue a different approach by posing the following question: If one is ignorant of the drift parameters, what is the best possible hedge?

One way of doing this is by assuming that $\rho=1\rho=1$, in which case no-arbitrage considerations imply that $\theta U=\theta S$. In this case we obtain (6) as an approximate hedging strategy, which is independent of the drift coefficients of the assets. Monoyios [5] uses this strategy as a benchmark in his numerical experiments (even when $\rho\neq1\rho\neq1$). Using the capital asset pricing model (CAPM) as inspiration, we shall now derive an improved benchmark strategy, which is independent of the drifts of the underlying assets, without imposing any assumptions on the value of ρ.

Under the assumptions of the CAPM, we can express the "beta" of U, with respect to the "market portfolio" S, as follows:

$$\beta_U := \rho\frac{\sigma_U}{\sigma_S}.$$

The following relation then expresses the drift rate of U in terms of the drift rate of S, under the assumptions of CAPM equilibrium:

$$\mu_U - r = \beta_U(\mu_S - r)$$

(see, e.g., Luenberger [24], §7.3, p. 177). The above two equations imply that

From (11) it now follows that $\kappa=0\kappa=0$, and therefore (17) yields the following naive approximation to the local risk-minimizing strategy:

$$\widehat{\xi}_t^{\mathsf{n}} := \rho \frac{\sigma_U U_t}{\sigma_S S_t} \Delta_{\mathsf{BS}}(t, U_t, 0, \sigma_U). \tag{22}$$

Note that this strategy requires no knowledge of the drift coefficients of the assets, but nevertheless imposes no constraints on the correlation coefficient. In Section 9, where it is used as a benchmark for our numerical simulations, we shall see that it performs substantially better than the naive strategy of Monoyios [5] outlined above.

Of course, it is possible to raise numerous objections to the CAPM assumptions in the context above. Nevertheless, as we shall see in the results of our numerical simulations, the hedging strategy (22) performs remarkably well. Thus, even if the CAPM is economically unjustified in our setup, taking account of the actual value of ρ improves hedging performance.

9. HEDGE SIMULATION RESULTS

To evaluate the effectiveness of hedging using the quadratic techniques, we now analyze the results of some hedge simulations. Initially, a comparison of the quadratic techniques with the approximate schemes obtained by Monoyios [5,6] was undertaken for a European put option. The put was written on the non-traded asset U and the risk was hedged by trading in S. Table 1 lists the model parameters used.

Table 1. Model parameters employed by the hedge simulations (the same parameters as used by Monoyios [5]).

U_0	S_0	K	r	μ_U	σ_U	μ_S	σ_S	T
100	100	100	5%	0.12	0.30	0.10	0.25	1 year

A Monte Carlo experiment was undertaken to test hedging performance. One million paths for U and S were generated, and rebalancing was allowed to take place 200 times, at equal intervals, over the life of the option. The approximation price of the option was used as the initial endowment. At the end of the period, the difference between the accumulated gain from hedging and the expiry value of the option was recorded as a profit or loss.

Five strategies were used to compute the hedge parameters. These were the naive strategy (22), the local risk-minimizing hedge ratio (17), the mean-variance optimal hedge ratio (18), the O(a4) hedge ratio proposed in Section 4.1.1 of Monoyios [5], and the O(a5) hedge ratio in Proposition 2 of Monoyios [6].

Two values of ρ were used, namely 0.85 and 0.95, while three values of the risk aversion constant γ were used for the Monoyios algorithms, namely 0.001, and 0.1. Tables Table 2 and Table 3 provide summary statistics for the Monte Carlo experiments.

It should be noted that the O(a4) and O(a5) Monoyios algorithms give almost identical results for γ=0.001 and γ=0.01. This is due to the fact that the parameter a is very small for these values of γ, making the fifth order term in the expansions (20) and (21) irrelevant—see the discussion in Section 7. The fifth order term only results in a difference when γ=0.1γ=0.1, and even then, only for the smaller correlation coefficient.

Histograms of the resulting terminal profits or losses are given in Figure 1. Since the γ=0.001 results are not significantly different from those of the local risk-minimizing strategy, and since the γ=0.1 histograms are not very accurate, we only provide histograms for the O(a5) Monoyios algorithm with γ=0.01.

Table 2. Summary statistics of hedging error for the put option with parameters given by Table 1 and ρ=0.85.

Strategy	Max	Min	Mean	SD	Median
Naive	36.02	−44.99	−0.0341	6.6530	0.9150
Local Risk	35.49	−44.94	0.0123	6.6487	1.1130
Mean variance	40.64	−47.24	0.0113	6.5908	1.0246
Monoyios $O(a^4), \gamma = 0.001$	35.52	−44.94	0.0101	6.6487	1.1037
Monoyios $O(a^5), \gamma = 0.001$	35.52	−44.94	0.0101	6.6487	1.1037
Monoyios $O(a^4), \gamma = 0.01$	35.78	−44.96	−0.0096	6.6498	1.0231
Monoyios $O(a^5), \gamma = 0.01$	35.78	−44.96	−0.0096	6.6498	1.0231
Monoyios $O(a^4), \gamma = 0.1$	37.90	−44.95	−0.2101	6.7894	0.2264
Monoyios $O(a^5), \gamma = 0.1$	37.90	−44.95	−0.2094	6.7880	0.2299

Table 3. Summary statistics of hedging error for the put option with parameters given by Table 1 and ρ=0.95.

Strategy	Max	Min	Mean	SD	Median
Naive	23.96	−28.98	−0.0283	4.0133	0.2169
Local Risk	23.59	−29.11	0.0072	4.0075	0.3554
Mean variance	26.08	−30.76	0.0064	3.9726	0.3249
Monoyios $O(a^4)$, $\gamma = 0.001$	23.60	−29.11	0.0063	4.0075	0.3521
Monoyios $O(a^5)$, $\gamma = 0.001$	23.60	−29.11	0.0063	4.0075	0.3521
Monoyios $O(a^4)$, $\gamma = 0.01$	23.70	−29.08	−0.0014	4.0079	0.3230
Monoyios $O(a^5)$, $\gamma = 0.01$	23.70	−29.08	−0.0014	4.0079	0.3230
Monoyios $O(a^4)$, $\gamma = 0.1$	24.71	−28.77	−0.0796	4.0453	0.0315
Monoyios $O(a^5)$, $\gamma = 0.1$	24.71	−28.77	−0.0796	4.0453	0.0316

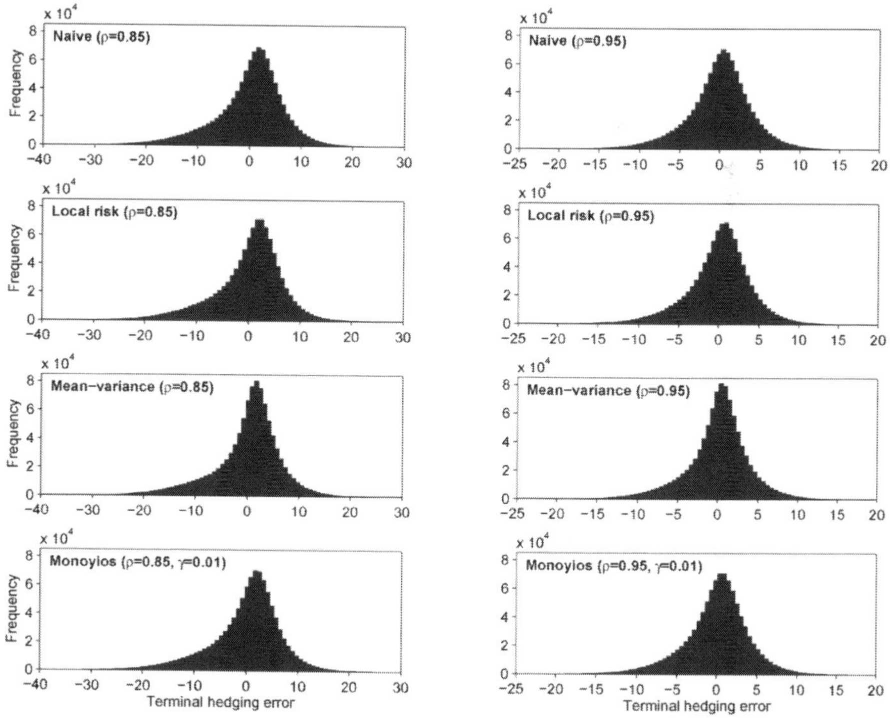

Figure 1. Histograms of the hedging errors for the put option, based on one million sample paths. The approximation prices were 8.6564 and 8.8733, corresponding to correlation coefficients of 0.85 and 0.95, respectively.

The results are encouraging, with the local risk-minimizing strategy performing at least as well as the Monoyios algorithm (*i.e.*, higher mean/median profit and lower standard deviation). This is not surprising, since for small values of γ the utility formulation is close to the local risk-minimizing strategy. The mean-variance optimal strategy performed slightly better than the other two, with a standard deviation that was 1%–2% lower. This can be seen as an enhanced peak around the mean in the relevant histograms. One slight drawback of this method is that its largest losses exceeded the largest losses of the other methods. The naive strategy performed surprisingly well (certainly significantly better than the naive strategy in Monoyios' simulations). This is due to the fact that, under the choice of parameters used, the CAPM equilibrium condition is not unreasonable.

Table Table 4 shows the approximation prices for various values of the correlation coefficient, for both put and call options, based on the parameters presented in Table 1. It is interesting to note that the approximation prices for the put are lower than the Black–Scholes prices, while the converse is true for the call. One should not interpret these prices as the premiums charged for the options, since not all risk is hedged, due to incompleteness. It is therefore necessary to estimate the standard deviation of the hedging error, so that the option writer can charge an appropriate risk premium.

Table 4. Put and call option approximation prices for various values of ρ. (For ρ=1ρ=1, they are Black–Scholes prices.)

ρ	**Put**	**Call**
−0.95	5.3127	23.7315
−0.75	5.6321	22.6965
−0.50	6.0493	21.4435
−0.25	6.4870	20.2358
0	6.9451	19.0730
0.25	7.4238	17.9549
0.50	7.9231	16.8812
0.75	8.4428	15.8514
0.95	8.8733	15.0588
1	9.3542	14.2312

Figure 2 shows the approximation prices and standard deviations of hedging errors for the put, using the naive, local risk-minimizing and mean-variance optimal hedging strategies. Figure 3 shows the same results for the call. The standard deviations were estimated based on Monte Carlo samples of 100,000 paths. We see from these graphs that hedging basis risk is only viable when the assets are highly correlated (in absolute value), since the error in hedging increases rapidly as the correlation between the assets decreases. It is interesting to note that the naive strategy performs very well for correlations close to unity. As the correlation coefficient reduces and becomes negative, it becomes less and less effective, however. This is due to the fact that, under the asset parameters used, the CAPM equilibrium assumption becomes less realistic as the correlation decreases.

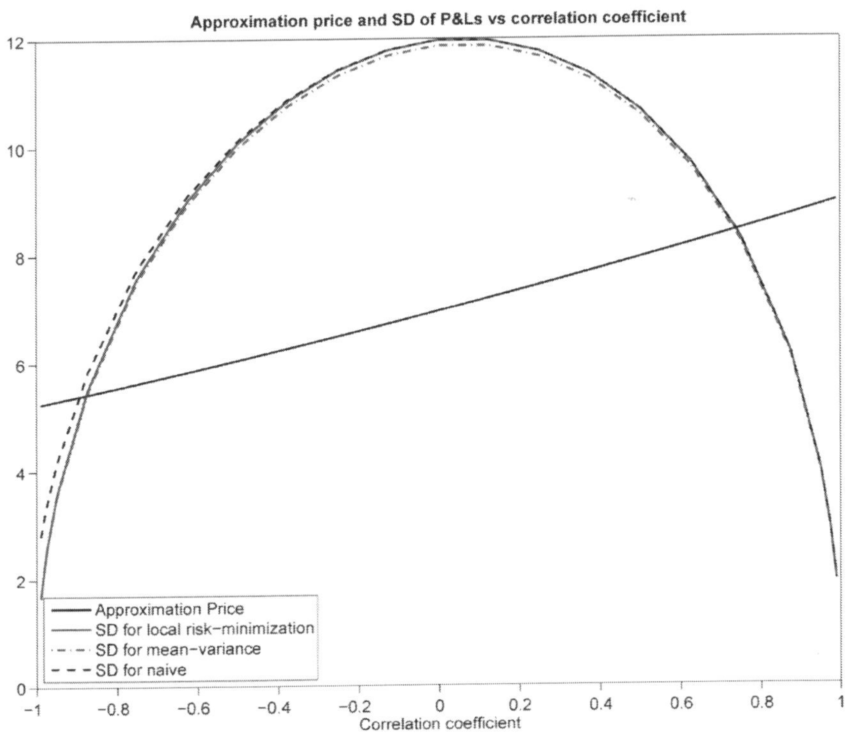

Figure 2. Approximation price and standard deviation of hedging error *vs.* correlation, for the put option with parameters given by Table 1.

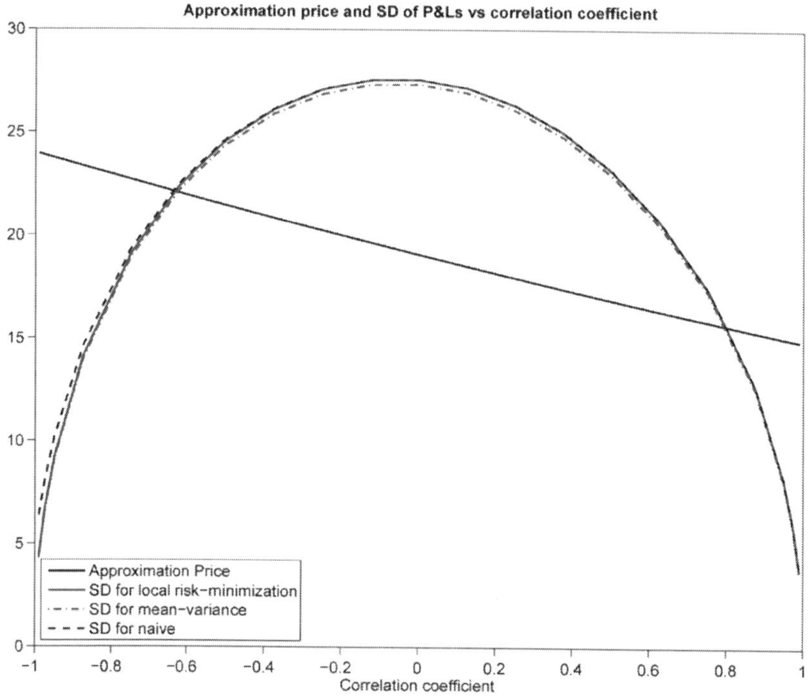

Figure 3. Approximation price and standard deviation of hedging error *vs.* correlation, for the call option with parameters given by Table 1.

10. CONCLUSIONS

In this paper we used quadratic criteria to derive simple hedging rules for minimizing basis risk. These rules are considerably simpler than the hedging rules based on utility maximization and perform slightly better in terms of minimizing risk. Their simplicity is a consequence of the fact that they are based on the Black–Scholes formula. By contrast, the utility maximization approach requires the use of series expansions in order to solve the relevant PDEs, which were originally derived using a "distortion" technique. It is important to note that all hedging schemes are only effective when the traded and non-traded assets are highly correlated (in absolute value), thus confirming the sobering conclusions of Davis [2].

As is usual in incomplete market settings, the solutions depend on estimating the growth rates of the assets—a task recognized to be very difficult. In order to address this issue, we have derived a naive hedging strategy that does not depend on the drift parameters. However, this comes with an implicit drawback—the performance of the strategy is dependent on how well a CAPM equilibrium condition is obeyed. In order to establish if this assumption is reasonable, in the context of specific real-world applications, further investigation is required.

ACKNOWLEDGMENTS

The authors would like to thank three anonymous reviewers of this paper for their helpful comments and revisions.

AUTHOR CONTRIBUTIONS

T.M. conceived the problem, performed the analysis and conducted computational experiments; H.H. and T.M. wrote the paper.

CONFLICTS OF INTEREST

The authors declare no conflicts of interest.

REFERENCES

1. Davis, M.H.A. Option Valuation and Hedging with Basis Risk. In *System Theory: Modeling, Analysis and Control*; Djaferis, T.E., Schuck, I.C., Eds.; Kluwer: New York, NY, USA, 1999; pp. 245–254.
2. Davis, M.H.A. Option Hedging with Basis Risk. In *From Stochastic Calculus to Mathematical Finance*; Kabanov, Y., Liptser, R., Stoyanov, J., Eds.; Springer: Berlin, Germany, 2006; pp. 169–187.

3. Henderson, V. Valuation of Claims on Nontraded Assets Using Utility Maximization. *Math. Financ.* **2002**, *12*, 351–373.

4. Henderson, V.; Hobson, D.G. Substitute Hedging. *Risk* **2002**, *15*, 71–75.

5. Monoyios, M. Performance of Utility-Based Strategies for Hedging Basis Risk. *Quant. Financ.* **2004**, *4*, 245–255.

6. Monoyios, M. Optimal hedging and parameter uncertainty. *IMA J. Manag. Math.* **2007**, *18*, 331–351.

7. Zariphopoulou, T. A solution approach to valuation with unhedgeable risks. *Financ. Stoch.* **2001**, *5*, 61–82.

8. Schweizer, M. Mean-Variance Hedging for General Claims. *Ann. Appl. Probab.* **1992**, *2*, 171–179.

9. Duffie, D.; Richardson, H.R. Mean-variance Hedging in Continuous Time. *Ann. Appl. Probab.* **1991**, *1*, 1–15.

10. Pham, H. On Quadratic Hedging in Continuous Time. *Math. Methods Oper. Res.* **2000**, *51*, 315–339

11. Schweizer, M. A Guided Tour through Quadratic Hedging Approaches. In *Option Pricing, Interest Rates and Risk Management*; Jouini, E., Cvitanić, J., Musiela, M., Eds.; Cambridge University Press: Cambridge, UK, 2001; Chapter 15; pp. 538–574.

12. McWalter, T.A. Quadratic Criteria for Optimal Martingale Measures in Incomplete Markets. *M.Sc. Dissertation*; University of the Witwatersrand: Witwatersrand, South Africa, 2006. http://hdl.handle.net/10539/2084 (accessed on 29 January 2015).

13. Harrison, J.M.; Pliska, S.R. Martingales and Stochastic Integrals in the Theory of Continuous Trading. *Stoch. Process. Their Appl.* **1981**, *11*, 215–260.

14. Harrison, J.M.; Pliska, S.R. A Stochastic Calculus Model of Continuous Trading: Complete Markets. *Stoch. Process. Their Appl.* **1983**, *15*, 313–316.

15. Schweizer, M. Risk-minimality and orthogonality of martingales. *Stoch. Stoch. Rep.* **1990**, *30*, 123–131.

16. Schweizer, M. Option hedging for Semimartingales. *Stoch. Process. Their Appl.* **1991**, *37*, 339–363.

17. Föllmer, H.; Schweizer, M. Hedging of Contingent Claims Under Incomplete Information. In *Applied Stochastic Analysis*; Davis,

M.H.A., Elliott, R.J., Eds.; Gorden and Breach Science Publishers: New York, NY, USA, 1991; pp. 389–414.

18. Schweizer, M. On the minimal martingale measure and the Föllmer-Schweizer decomposition. *Stoch. Anal. Appl.* **1995**,*13*, 573–599.

19. Heath, D.; Platen, E.; Schweizer, M. A Comparison of Two Quadratic Approaches to Hedging in Incomplete Markets.*Math. Financ.* **2001**, *11*, 385–413.

20. El Karoui, N.; Peng, S.; Quenez, M.C. Backward Stochastic Differential Equations in Finance. *Math. Financ.* **1997**, *7*, 1–71.

21. Shreve, S.E. *Stochastic Calculus for Finance II: Continuous-Time Models*; Springer-Verlag: New York, NY, USA, 2004.

22. Rogers, L.C.G. The relaxed investor and parameter uncertainty. *Financ. Stoch.* **2001**, *5*, 131–154.

23. Monoyios, M. Utility-based valuation and hedging of basis risk with partial information. *Appl. Math. Financ.* **2010**, *17*, 519–551.

24. Luenberger, D.G. *Investment Science*; Oxford University Press: New York, NY, USA, 1998.

5

CHAPTER

Predicting Conditional Autoregressive Value-at-Risk for Stock Markets during Tranquil and Turbulent Periods

Anastassios A. Drakos[1], Georgios P. Kouretas[2*], Leonidas Zarangas[3]

[1]*Department of Business Administration, Athens University of Economics and Business, Athens, Greece*

[2]*IPAG Business School, France and Department of Business Administration, Athens University of Economics and Business, Athens, Greece*

[3]*Department of Finance & Auditing, Technological Educational Institute of Epirus, Preveza, Greece*

ABSTRACT

This paper analyzes the predictive performance of the Conditional Autoregressive Value at Risk (CAViaR) developed by Engle & Manganelli (2004) for major equity markets during tranquil and turbulent periods. The CAViaR model shifts the focus of attention from the distribution of returns directly to the behaviour of the quantile. We compare the predictive performance of four alternative CAViaR specifications, namely Adaptive, Symmetric Absolute Value, Asymmetric Slope and Indirect GARCH(1,1) models due toEngle & Manganelli (2004) along with the improved asymmetric CAViaR (I-CAViaR) model due to Huang et al. (2009). We employ daily returns for six stock markets indices, namely S & P500, FTSE100, NIKKEI225, DAX30, CAC40 and Athens Exchange General index for the period January 2, 1995 to August 23, 2013. We compare the predictive performance of the alternative specifications for three subperiods: before, during and after the recent 2007-2009 financial crisis. The comparison is done with the use of a battery of tests which includes unconditional and conditional coverage tests, the Dynamic Quantile high-order independence test and the White (2000) empirical coverage probability and predictive quantile loss tests. The main findings of the present analysis is that the CAViaR quantile regression models and the I-CAViaR model have shown significant success in predicting the VaR measure for various periods although this performance varies over the three periods before, during and after the 2007-2009 financial crisis.

Keywords: Non-Linear Regression Quantile, CAViaR, VaR, Quantile Loss, Reality Check, Backtesting

1. INTRODUCTION

The global financial crisis of 2007-2009 has called once again into question financial risk management practices, and one key issue is whether risk measures can actually be forecasted accurately enough to accomplish this task. Thus the current financial crisis has indeed highlighted once again the importance of risk management where institutions such as banks and major hedge fund companies are found to inevitably fail from time to time as we have witnessed the last twenty years. Quantitative risk measure forecasting has become, at least since the market crash in 1987 and then as a consequence of the financial crisis of 1997-1998, as well as the bankruptcy of several financial institutions such as the BCCI and Barings international banks that led to increased price volatility and financial uncertainty the benchmark for measuring market risk. Such financial uncertainty has increased the likelihood of financial institutions suffering substantial losses as a result of their exposure to unpredictable market changes. These events have made investors become more cautious in their investment decisions, while it has also led to an increased need for more careful study of price volatility in stock markets. Moreover, the recent crisis has shown the necessity of having adequate risk-management protocols in order to achieve greater resilience, and hence, the need to improve the existing procedures for quantifying the market risk. The present paper is motivated mainly by this concern.

In bank regulation, the effectiveness of capital requirements in preventing funding shortfall rests upon the estimation accuracy of market risk measures. Market risk is one of the four types of risk that financial institutions can expose them to. It is considered the most significant one since it represents potential economic loss caused by the reduction in the market value of a portfolio. The existence of market risk and recent financial disasters has raised the need for the development of practical risk management tools for financial institutions. This need had been reinforced by the Basel Committee of Banking Supervision (1996) which called for the use of internal market risk management to capital requirement by financial institutions such as banks and investment firms.[1]The Basel II Accord is designed to monitor and encourage sensible risk-taking using appropriate models to calculate VaR and daily capital changes. For a financial institution, measuring the market risk it faces is an essential task. Specifically, a possible method of measurement is the evaluation of losses likely to be incurred when the price of the portfolio's assets declines.

Value-at-Risk has become the standard tool used by financial analysts to measure market risk. VaR is defined as a certain amount lost on a portfolio of financial assets with a given probability over a fixed number of days. The confidence level represents "extreme market conditions" with a probability that is usually taken to be 99% or 95%. This implies that only 1% (5%) of the cases will lose more than the reported VaR of a specific portfolio. VaR is widely used because of its simplicity. Essentially, the VaR provides a single number that represents market risk and therefore it is easily understood.[2]

Although the VaR is conceptually a simple measure of market risk, there exists a controversy with respect to the suitability of the alternative existing techniques employed to estimate the VaR. Indeed, the measurement of VaR is a very interesting statistical problem. Artzner et al. (1997, 1999) have derived a set of axioms that specify a coherent risk measure. Thus, a risk measure must possess the following characteristics. First, it should not exceed the maximum possible loss which can occur. Second, the proposed risk measure should be greater than the mean loss, implying capital adequacy to cover losses. Third, in the event that there is a proportional change in the loss, then we require the risk measure to change proportionally as well. Finally, it must satisfy the property of superadditivity, implying that the risk measure calculated for two separate losses should be equal to the risk measure calculated on the sum of the two portfolios. As Boyle et al. (2005), Alexander et al. (2006) and Longin (2001) among others emphasized, the VaR methodology had certain limitations since it did not satisfy the properties of subadditivity and excess of the mean loss. Given these reservations regarding the use of the VaR as a measure for market risk, several researchers have developed alternative risk measures.[3]

Calculating the VaR requires accurate knowledge of the distribution of extreme events. This is a difficult task since the distribution of portfolio returns is not constant over time and, given that VaR is nothing more than a specific quantile of future portfolio values subject to current information, we must find an appropriate model for time varying conditional quantiles.

This crucial issue is coupled with the need for providing accurate estimates of the chosen distribution of portfolio returns.

During the last two decades a large number of alternative models have been developed to estimate VaR. These alternative methodologies have mainly focused on modeling the entire distribution of returns and are based on the strict assumptions of normality or i.i.d. returns. However, as Engle & Manganelli (2004) argue, if we do not correctly estimate the underlying market risk then this can lead to an allocation of capital below first-best and that can affect the profitability and/or the financial stability of the corresponding bank or investment firm. Engle & Manganelli (2004) have proposed an alternative approach that models not the entire distribution but rather focuses on the regression quantile which does not require the above mentioned strict assumptions. This methodology, which is called Conditional Autoregressive Value at Risk (CAViaR), uses an autoregressive process in order to model the evolution of the regression quantile over time. The estimation of the unknown parameters is done with the use of the framework suggested by Koenker & Bassett (1978) . Furthermore, Engle & Manganelli (2004) prove that these estimators are asymptotically efficient and consistent. Finally, they develop the Dynamic Quantile test which is used to examine the quality of the CAViaR results.[4]

The main objective of this paper is to investigate the predictive performance of various types of the CAViaR specifications for stock market returns during tranquil and turbulent periods. The choice of this particular semiparametric method is motivated by the fact that CAViaR does not require any assumption on the distribution of a time series and computes the VaR directly by quantile regression, implying that the model allows the time series to switch from one stochastic process to another. We estimate and perform an evaluation of the predictive performance of two of the four alternative CAViaR specifications, namely, Symmetric Absolute Value and Asymmetric Slope. Furthermore, given the recent financial crisis we are further interested in evaluating the predictive performance of the alternative specifications for three out-of-sample evaluation periods (before-crisis, crisis and after-crisis). Within this framework we consider the issue of model stability. It is well documented that the model parameters are estimated from real data, which are often subject to structural changes due to regime shifts or events such as financial crises.

Recently, Huang et al. (2009) developed an improved CAViaR specification to account for some shortcomings of the original specification. Specifically, they argue that the asymmetric specification of Engle & Manganelli (2004) which is used to investigate the different effects of positive and negative returns on the VaR prediction requires the estimation of four parameters in the optimization procedure. This complicated model may lead to more estimation errors leading to an unstable or unrobust model.

Therefore, this extension of the constant-parameter CAViaR allows us to examine whether the parameters of an individual risky asset are driven by the market index return. This extended model allows the tails of a financial series to follow different stochastic process and it allows VaR to be influenced by the volatility level of the driving index.

The main objective of this paper is to investigate the predictive performance of various types of the conditional autoregressive value-at-risk (CAViaR) specification during tranquil and turbulence periods. To this end we employ daily returns for six stock markets indices, namely S & P500, FTSE100, NIKKEI225, DAX30, CAC40 and Athens Exchange General Index for the period January 2, 1995 to August 23, 2013. We compare the predictive performance of the alternative specifications for three subperiods: before, during and after the recent 2007-2009 financial crisis.

The main findings of the present analysis is that the CAViaR quantile regression models and the I-CAViaR model have shown significant success in predicting the VaR measure for various periods although this performance varies over the three periods before, during and after the 2007-2009 financial crisis.

The remainder of the paper is organized as follows. Section 2 presents some of the most widely used VaR models. In section 3 we discuss the CAViaR methodology and its proposed alternative specifications. In Section 4 we report our empirical results, and finally section 5 provides our concluding remarks.

2. VALUE-AT-RISK-MODELS AND METHODS

During the 1990s several alternative modeling methodologies for the estimation of the VaR were advanced. The purpose of these models was to provide risk managers with a comprehensive and intuitively easily understood measure of the VaR. The motivation for the development of the VaR models relies on the stylized characteristics of financial data which were first documented byMandelbrot (1963) and Fama (1965). To recapitulate, these characteristics imply that the returns of financial assets have leptokurtic distributions, that their distributions are negatively skewed and finally that they exhibit volatility clustering. As Manganelli & Engle (2004) and Engle & Manganelli (2004) point out, these alternative methodologies adopted a common general structure: (a) Mark- to-market the portfolio on a daily basis; (b) Estimation of the distribution of returns; and (c) Computation of the portfolio's VaR. The main difference among the alternative methodologies is linked to the estimation of the appropriate distribution of the portfolio returns. We can briefly discuss the advantages and disadvantages of the alternative VaR models using the following broad classification.[5]

The first class of models is fully parametric and includes applications such as J.P. Morgan'sRiskmetrics (1996) and GARCH models. These methodologies combine an econometric model with the assumption of conditional normality for the returns series. Specifically, these models rely on the specification of the variance equation of the portfolio returns and the assumption that the standardized errors are i.i.d. Additionally, when the GARCH methodology is applied we are also required to specify the distribution of the errors, which is usually taken to be the normal one, while it is assumed that the negative returns follow the same process as the rests of portfolio returns (Bams et al., 2005 ; Burns, 2005 ; Angelidis et al., 2004 ; Alexander et al., 2006 ; Pojarlev & Polasek, 2000 ; Polasek & Pojarlev, 2005 ; Kuester et al., 2006 ; Haas et al., 2006 ; Guidolin & Timmerman, 2006 ; Chen et al., 2013) are among the numerous applications employing alternative GARCH specifications. McAleer et al. (2008) develop a new Portfolio Spillover GARCH model and they examine its forecasting performance relative to that of the VARMA-GARCH with spillover effects and the CCC model with no spillovers. They find that spillovers effects are statistically significant.

The application of the parametric methodologies has been criticized since they tend to give coefficients which underestimate the VaR, mainly due to their failure to take into account the characteristic that the distribution of the portfolio returns has heavy tails. This underestimation of the VaR, as well as possible misspecifications with respect to the variance equation along with the distribution of errors, can be corrected by allowing alternative distributions of the errors such as the Gaussian, Student's-t and Generalized Error Distribution. However, it is further shown that the GARCH-type models provide satisfactory estimates of the quantile only when a bad event has already occurred.

The second approach for the estimation of the distribution of profits and losses is the non-parametric historical simulation. This methodology makes no assumption about the distribution of the portfolio returns and is based on the concept of rolling windows. The idea is to select a window which is usually taken to be anywhere between 6 months to 2 years and assume that any portfolio return has the same likelihood to occur. Moreover, a return which falls outside the chosen window has probability equal to zero to occur. This methodology has several deficiencies. It is inappropriate to provide extreme quantiles since we cannot extrapolate beyond past observations. The proposed solution to this problem is the increase of the sample of observations, but this will lead to estimates of the VaR which are biased downwards (or upwards) since we have a mixture of periods with low volatility with periods of high volatility.

Within this group of VaR models falls the hybrid approach developed by Boudoukh et al. (1998) , which combines the historical simulation and Riskmetrics. This methodology applies weights to the portfolio returns that decline exponentially. Although this approach improves on the previously discussed methodologies, it also has problems since the selection of the parameters as well the calculation of the VaR does not depend on sound statistical theory but is rather ad hoc.

The final group for the estimation of the VaR is the semiparametric models. The first approach in this category is the Extreme Value Theory proposed by Danielsson et al. (1998) and Danielsson & de Vries (2000) . The advantage of this approach is that it is based on sound statistical theory which offers a parametric form for the tail of a distribution. This approach focuses on the asymptotic form of the tail, rather than modeling the complete distribution of portfolio returns and therefore we are able to obtain more efficient forecasts of the risk associated with a particular market position. Although this methodology is very appealing it does have two shortcomings. First, as Danielsson & de Vries (2000) argue, this approach performs well at very low quantiles but fails to provide accurate estimations of the VaR at levels which are not considered very extreme. Second, this methodology is also based on the assumption of i.i.d. standardized errors which is, as we have already discussed, an important limitation. Despite its limitations this approach has found substantial applications recently (Longin, 2000 ; McNeil & Saladin, 2000 ; McNeil & Frey, 2000 ;Naftci, 2000 ; Bekiros & Georgoutsos, 2005a,b ; Brooks et al., 2005 ; Schaumburg, 2012).

Engle & Manganelli (2004) proposed an alternative semiparametric method to estimate VaR, namely Conditional Autoregressive Value at Risk (CAViaR). This approach is based on the simple intuition that it is better to model the quantile directly as it evolves through time instead of attempting to model and estimate the entire distribution of portfolio returns. Modelling the quantile instead of the entire distribution has the main advantage that we are not required to adopt the set of extreme assumptions which are invoked by alternative methodologies, among them normality or that returns are i.i.d. In addition, CAViaR takes into consideration volatility clustering of portfolio returns that leads to the understanding that the corresponding distributions are autocorrelated. As a consequence the VaR must also follow a similar pattern since it is directly linked with the standard deviation of the distribution. Therefore, Engle & Manganelli (2004) developed CAViaR in order to take into account of the particular characteristic of the VaR. Furthermore, Manganelli & Engle (2004) demonstrate that the CAViaR model outperforms most of the indirect-VAR methodologies. This outcome is reached through Monte Carlo simulations. They simulated 1000 samples of 2000 observations for seven different processes.

There are several papers that have extended and applied the CAViaR approach in a number of different ways. In an early study, Kouretas & Zarangas (2005) employ the CAViaR model to measure the market risk of five major equity markets, six blue chip stocks from the Athens Exchange, and six major company stocks listed on the New York Stock Exchange (NYSE). They compare four alternative CAViaR specifications with mixed results. Giacomini & Komunnjer (2005)find that CAViaR is most accurate at the 1% quantile level, but the GARCH model with Gaussian distributed errors is better than CAViaR at the 5% quantile level. Bao et al. (2006) compare the performance of CAViaR and other VaR measures for five East and Southeast Asian markets before, during and after the financial crisis of 1997-1998. They conclude that CAViaR models are quite satisfactory in a stable period but their performance in less satisfactory during turbulent periods.Kuester et al. (2006) compare CAViaR and alternative approaches for univariate VaR forecasts using daily return data on the NASDAQ Composite Index and alternative evaluation criteria. They conclude that CAViaR specifications performs well overall using unconditional and conditional tests for the predictive performance of VaR, although its forecasting performance is less accurate during periods of crisis and very high volatility.

Gourieroux & Jasiak (2008) argue that a drawback of the dynamic quantile models is the lack of an internal mechanism that would ensure that the conditional quantiles evaluated at different probability levels do not cross. They develop the Dynamic Additive Quantile model, which ensures the monotonicity of conditional quantile functions and using data for the Toronto Stock Exchange market index and it is shown the DAQ model provides better asymptotically efficient estimations.Huang et al. (2009) develop an extension of the CAViaR model (a) with a proposition of a new asymmetric CAViaR specification and (b) a mixed data regression model for multi- period VaR prediction. Using market data on WTI daily spot oil prices the improved CAViaR specifications performed well in a battery of evaluation criteria. Huang et al. (2010) introduce a time-varying CAViaR model and using data to construct size weighted portfolios from the NYSE, AMEX and NASDAQ they show that the time-varying model provides better VaR forecasts than the constant parameter CAViaR when there are spillover effects from one market segment to other markets or market segments. Yu et al. (2010) extend the CAViaR specifications using two approaches, namely the threshold and mixture type indirect-GARCH CAViaR models. Gerlach et al. (2011) also propose a nonlinear CAViaR model. Chen et al. (2012) adopt the Threshold CAViaR and the indirect CAViaR model and using data from major stock market indices they provide clear evidence that the CAViaR class of models outperform in forecasting VaR performance other competing parametric, semiparametric and nonparametric models. Jeon & Taylor (2013) proposed a model of CAViaR with implied volatility and they find that the implied volatility has more explanatory power that the standard VaR models. Finally, Rubia et al. (2013) analyze tail-predictability in the returns of several representative portfolios on the US stock exchange. They show that the CAViaR models are very suitable in accurate VaR forecasts when different variables which are related to market-wide measures of trading activity and liquidity.

3. CAVIAR

3.1. General Framework

Engle & Manganelli (2004) and Manganelli & Engle (2001) consider a vector of portfolio returns that is observable, defined as $\{y_t\}_{t=1}^{T}$. Let θ be the probability tied to VaR, x_t be a vector of observable variables at time t, and β_θ be a vector of unknown parameters. They also define $f_t(\beta) \equiv f(x_{t-1}, \beta_\theta)$ as the θ-quantile of the distribution of the portfolio returns at time t which has been formed at time $t-1$.[6]Therefore, a general formulation of CAViaR can be written as follows:

$$f_t(\beta) = \beta_0 + \sum_{i=1}^{q} \beta_i f_{t-i}(\beta) + \sum_{i=1}^{p} \beta_j l(x_{t-j}) \tag{1}$$

where $p = q + r + 1$ is the dimension of β and l is a function of a finite number of lagged values of observables. Moreover, in order for the quantile to have a smooth transition they use the autoregressive terms $\beta_i f_{t-i}(\beta), i = 1, \cdots, q$. Finally, they use the term $l(x_{t-j})$ to provide a relationship between the θ-quantile $f_t(\beta)$ and the observable variables which are included in the information set. As Engle & Manganelli (2004) point out, we can consider the lagged portfolio returns as the best choice for x_{t-1}. This implies that as y_{t-1} becomes negative, then one should expect the VaR to increase, while the VaR tends to decline in good days. Therefore, we expect that changes in y_{t-1} will symmetrically affect the VaR.

The purpose is to develop alternative specifications for the function l and then estimate the different models. Engle & Manganelli (2004) propose four alternative CAViaR specifications which we will estimate in our case.

3.2. Adaptive Specification

The first specification is called Adaptive and takes the following formulation:

$$f_t(\beta_1) = f_{t-1}(\beta_1) + \beta_1 \left\{ \left[1 + \exp\left(G\left[y_{t-1} - f_{t-1}(\beta_1) \right] \right) \right]^{-1} - \theta \right\} \quad (2)$$

where G is some positive finite number and we that as $G \rightarrow \infty$, the last term of equation (1) converges to $\beta_1 \left[I\left(y_{t-1} \leq f_{t-1}(\beta_1) \right) - \theta \right]$, where $I(.)$ is the indicator function. The intuition behind the adaptive specification tells us that in those cases where the VaR has been exceeded, then we should increase its value, whereas in those cases that we do not exceed it, then we should reduce its value by a small magnitude. Such a strategy will lead in a reduction in the probability of observing a sequence of hits while at the same time it is highly unlikely that we will have zero number of hits. Engle & Manganelli (2004) also point out that this CAViaR specification has a unit coefficient on the lagged VaR.

3.3. Symmetric Absolute Value Specification

A second specification is called Symmetric Absolute Value (SAV) and its mathematical formulation is given by:

$$f_t(\beta) = \beta_1 + \beta_2 f_{t-1}(\beta) + \beta_3 \left| y_{t-1} \right| \quad (3)$$

This model responds symmetrically to past portfolio returns and it is mean reverting since the coefficient of the lagged VaR is not constrained to equal one. Furthermore, we could properly specify this quantile specification using a GARCH model with the standard deviation (and not the variance), which is considered to follow a symmetric distribution with i.i.d. errors.[7]

3.4. Asymmetric Slope Specification

The Asymmetric Slope (AS) is the third commonly used specification to estimate the l function. It is written as follows:

$$f_t(\beta) = \beta_1 + \beta_2 f_{t-1}(\beta) + \beta_3 (y_{t-1})^+ + \beta_4 (y_{t-1})^- \qquad (4)$$

The Asymmetric Slope model allows for an asymmetric response to positive and negative past portfolio returns.[8] Again this model is mean reverting. As with the SAV model we can correctly specify this specification by fitting a GARCH process with the standard deviation, following this time an asymmetric distribution with i.i.d. errors.

3.5. Indirect GARCH(1,1)

The final specification is called Indirect GARCH (1,1) which is also mean reverting and as with the SAV specification, it responds symmetrically to past returns. This specification can be correctly modeled under the assumption that the underlying data process follows a true GARCH(1,1) with an i.i.d. error distribution.[9] The algebraic expression of this specification is as follows:

$$f_t(\beta) = \left(\beta_1 + \beta_2 f_{t-1}^2(\beta) + \beta_3 y_{t-1}^2\right)^{1/2} \qquad (5)$$

Thenext step to the analysis is the estimation of the parameters of the alternative CAViaR models. They are estimated using linear and non-linear quantile techniques. These techniques were first introduced by Koenker & Basett (1978) who provide a thorough analysis of how to apply the concept of sample quantile to a linear regression model. We consider the following model proposed by Engle & Manganelli (2004) :

$$y_t = x_t'\beta^0 + \varepsilon_{\theta t},\ Quant_\theta\left(\varepsilon_{\theta t}|x_t\right) = 0 \tag{6}$$

where x_t is a p-vector of regressors and $Quant_\theta\left(\varepsilon_{\theta t}|x_t\right)$ is, as we have already defined, the θ-quantile of $\varepsilon_{\theta t}$ conditional on x_t. White (1994) has shown that if we minimize the regression quantile objective function that was developed by Koenker & Basett (1978) we can obtain consistent estimates under certain assumptions. This minimization can be considered as follows: We define $f_t(\beta) \equiv x_t\beta$. Then any $\hat{\beta}$ that solves the following problem:

$$\min_{\beta} \frac{1}{T}\left[\theta - I\left(y_t < f_t(\beta)\right)\right]\left[y_t - f_t(\beta)\right] \tag{7}$$

defines the θth regression quantile.

Within this framework, Engle & Manganelli (2004) show that the only assumption required is the appropriate specification of the quantile process and, more specifically, we do not have to specify the entire distribution of the error terms. Furthermore, even if we erroneously specify the regression quantile process, Engle & Manganelli (2004) argue that we can still obtain a minimization of Equation (5) that satisfies the Kullback-Leibler Information Criterion (which measures the deviation between the true specification and the actual model).

3.6. Improved CAViaR

As we already discussed Huang et al. (2009) argue that in practice CAViaR may face some implementation challenges. Thus, in case we investigate the different effects of positive and negative returns on the prediction from a VaR model then we may use the asymmetric CAViaR specification. This requires estimating four parameters in the optimization framework which may lead to more estimation errors and this could result to an unstable or unrobust model. Based on the exponentially weighted moving average volatility prediction idea in Riskmetrics (1996), Hull & White (1998), and Huang et al. (2009) they propose an improved asymmetric CAViaR model:

$$q_t(\theta) = \alpha + \beta_0 q_{t-1}(\theta) + (1-\beta_0)\left(\frac{v}{1-\beta_1} I(y_{t-1} > 0) + \frac{v}{\beta_1} I(y_{t-1} < 0)\right)\left|y_{t-1} - u\right|$$

(8)

where $v = \sqrt{\beta_1^2 + (1-\beta_1)^2}$, $(0 < \beta_1 < 1)$ and u is the sample mean. This CAViaR specification allows the VaR

prediction to respond asymmetrically to positive and negative returns. Thus, if we consider that $\beta_1 < 0.5$, negative returns have a larger influence on the VaR prediction. In addition, in contrast to the asymmetric specification (4) this improved asymmetric specification has only three parameters to be estimated. Furthermore, Huang et al. (2009) argue that the Improved CAViaR model embodies two important characteristics. First, it can capture skewness in the stock market indices returns. Second, the model is consistent with the leverage effect. That is, changes in volatility are negatively correlated with changes in prices and this is a feature that is directly applied to the VaR prediction procedure.

Engle & Manganelli (2004) consider the case where $\hat{\beta}$ is a non-linear regression quantile estimator and they prove that this estimator is consistent and asymptotically normal. Furthermore, they show that there is a consistent estimator of the variance-covariance matrix. Engle & Manganelli (2004)then go on to derive the asymptotic distribution of the estimator. This allows us to conduct hypothesis tests of the quantile models.[10]

3.7. Evaluation Performance Tests

Engle & Manganelli (2004) also propose a new test for the evaluation of the alternative specifications which has better power properties than other existing tests. This test allows for the inclusion of a variety of alternative specifications. They define:

$$Hit_t(\beta^0) \equiv f\left(y_t < f_t(\beta^0)\right) - \theta \tag{8}$$

where the function $Hit_t(\beta^0)$ is assumed to take a value $(1-\theta)$ every time y_t falls below the quantile, and it takes the value $-\theta$ in all other cases. Equation (8) implies that the expectation of $Hit_t(\beta^0)$ is zero. Furthermore, based on the definition of the quantile given in equation (1), we also assume that the conditional expectation of $Hit_t(\beta^0)$, given a set of information at period $t-1$, is zero. This implies that $Hit_t(\beta^0)$ must be uncorrelated with its own lagged values as well as with $f_t(\beta^0)$ and its expected value should equal zero. If these assumptions hold for $Hit_t(\beta^0)$ then we are certain that we have no misspecification error introduced, there is no autocorrelation in the hits, and we will obtain the correct fraction of exceptions. Based on definition (8), Engle & Manganelli (2004) derive two test statistics. First, they construct an in-sample Dynamic Quantile test. This test is a specification test which is used to select among alternative model specifications of a particular CAViaR process. Second, they construct an out-of-sample Dynamic Quantile test. This test is useful to the market regulators and/or the risk managers, since they can examine whether the VaR estimates for a particular financial institution satisfy certain properties such as that they are unbiased, they provide ind ependent hits and they give quantile estimates which are ind ependent. Moreover, it is argued that this second test has some nice features since it is simple in its application and it does not depend on the procedure used for the estimation. We obtain the results from this test simply by using a series of VaR values and the respective value of the portfolio.[11]

An important issue when we compare alternative models in terms of predictive performance using the same data, is the need to take into account the dependence among the models. White (2000)developed a test of superior unconditional predictive ability among multiple models accounting for specification search. Furthermore, we note that when we apply backtesting procedure we may be interested not only to the behaviour of hit sequence (i.e. whether an exceedance occurs) but also in the magnitude of the exceedance. In general, a backtest could be based on a function of the observed return series and the corresponding VaR predictions. This would imply the construction of a general loss function which could be evaluated using past data on returns and the reported VaR series. Following Bao et al. (2006) and Huang et al. (2009) define the loss function as

$$Q_t(\theta)=\left[\theta-H_t(\theta)\right]\left[r_t+q_t(\theta)\right] \tag{9}$$

The null hypothesis is that the empirical value of $Q_t(\theta)$ for one model is less than or equal to a competing one, which implies that the competing model cannot outperform the first model. To compute the p-value of the hypothesis, we adopt the statistics developed by White (2000), corresponding to the empirical loss $Q(\theta)$ and the reality check statistic where $H_t(\theta)$ denotes the previous value of $Hit_t\left(\beta^0\right)$.[12]

4. EMPIRICAL RESULTS

We apply the alternative CAViaR model specifications on daily data for the period March 3, 1995 to August 28, 2013 for the following general stock indices: S & P500, FTSE100, NIKKEI225, DAX30, CAC40 and the Athens Exchange General Index. The data was taken from Datastream. Given our earlier discussion that financial returns exhibit certain properties which imply that it is unreasonable to assume that stock return series follow a certain stochastic process. This evidence motivates this study to consider the CAViaR model by accommodating different processes into the prediction horizon. Table 1 reports summary statistics for each stock market daily returns. The significant negative skewness and kurtosis show that the return process deviates from the normal distribution. The daily returns are computed as 100 times the difference of the log of the prices. Finally in order to implement our analysis we construct historical series of portfolio for each case and we choose a specification of the functional form of the quantile.

Our analysis begins with estimation of the Symmetric, Asymmetric and the Indirect GARCH (1,1) CAViaR specifications due to Engle & Manganelli (2004) and the Improved Asymmetric CAViaR due to Huang et al. (2009) described in Section 3. For the estimation of the models we used the first 4364 observations and the last 500 to conduct the out-of-sample testing performance.

We estimate 1% and 5% one day Value-at-Risk.[13] As Engle & Manganelli (2004) proved, all the models are both continuous and continuously differentiable in β.[14] Given that the model parameters are estimated from data, which in many cases are subject to structural breaks we estimate the four alternative CAViaR specifications for three subsamples. The first subsample covers the pre-crisis period 3 January 1995 to 14 September 2007, the second subsample covers the crisis period 17 September to 31 August 2009 and the final subsample refers to the post-crisis period 1 September 2009 to 23 August 2013.

Our results are summarized in Tables 1-6. Each table reports the value of the estimated parameters, the respective standard errors and the one-sided p-values. Furthermore, each table shows the value of the regression objective function given by equation (3) above. Finally, we report the percentage of times the VaR is exceeded and the in-and out-of-sample p-value of the Dynamic Quantile test. The computation of the VaR series with the CAViaR models has been done with the initialization of $f_1(\beta)$ to the empirical θ-quantile of the first 500 observations. With respect to the computation of the DQ test we used a constant, the VaR forecast and the first four lagged hits as instrumental variables. In contrast, to avoid the presence of collinearity in the matrix of the first and higher order derivatives, we did not include the constant and the VaR forecast.[15] Following Engle & Manganelli (2004) we compute the standard errors and the variance covariance matrix of the in-sample DQ test and the calculation of the statistics $\hat{D}_T$ and $\hat{M}_T$ was done with the use of the k-nearest neighbor estimators, with $k = 40$ for the 1% VaR and $k = 60$ for the 5% VaR.[16]

The results obtained for the pre-crisis period are reported in Table 1 and Table 2. The first important observation we make in both tables is that the coefficient β_2 is very significant and this implies that volatility clustering is verified for the stock price returns of the six general indices. More specifically, this carries over to the tails of the distribution. Second, we note the accuracy of the alternative models. This is measured by the percentage of in-sample hits. Consider the results of the 1% VaR. We observe that for either the case of the mature market or the emerging market, or even the case of the general stock indices the Symmetric Absolute Value, the Asymmetric Slope and the Indirect GARCH models provide estimates which are extremely close to the value of 1, which is taken as evidence that they describe the evolution of the tail for most of the cases. Specifically, the results are particularly good for the stock returns of the S & P 500, FTSE100 and DAX30. In these cases we further observe that the out-of-sample hits are exactly equal to one or about 1.2%. For the stock returns of CAC40, NIKKEI225 and ASE the accuracy of the in-sample hits is fairly good, but the out-of sample hits are substantially below the value of 1. Furthermore, we see that the Symmetric Absolute Value, the Asymmetric Slope and the Indirect GARCH models provide accurate in-sample estimations, whereas the I-CAViaR specifi cation provides the best estimates at the 1% benchmark. Furthermore, the Symmetric Absolute Value, the Asymmetric Slope and the Indirect GARCH provide in-sample hits near the value of 1, but the out-of sample hits are some distance from this value. This is less problematic for the case of the I-CAViaR specification.

Table 1. Estimates and Relevant Statistics for the four Conditional Autoregressive Value at Risk Models-Pre Crisis period (03/01/1995-14/9/2007).

1% VaR	Symmetric Absolute Value			Asymmetric slope			Indirect GARCH			Improved Asymmetric		
	CAC40	DAX30	ASE	CAC40	DAX30	ASE	CAC40	DAX30	ASE	CAC40	DAX30	ASE
Beta 1	**0.1078**	**0.1574**	**−0.0013**	0.1318	**0.1502**	**0.0595**	0.2109	0.4298	−0.0001	**0.9742**	**0.1167**	**0.9398**
Standard Errors	0.0740	0.0597	0.0053	0.0428	0.0692	0.2812	0.1213	0.1705	0.0658	0.0664	0.0074	0.2005
p values	0.0725	0.0042	0.0039	0.0010	0.0151	0.0172	0.0410	0.0058	0.4962	0.0000	0.0000	0.0006
Beta 2	**0.9134**	**0.8785**	**0.9791**	**0.9062**	**0.8659**	**0.6712**	**0.9311**	**0.8657**	**0.9582**	0	0	0
Standard Errors	0.0412	0.0309	0.0050	0.0309	0.0541	0.0790	0.0136	0.0156	0.0091	0	0	0
p values	0.0000	0.0000	0.0000	0.0000	0.0000	0.0000	0.0000	0.0000	0.0000	0	0	0
Beta 3	**0.1746**	**0.2687**	**0.0980**	0.0945	**0.1988**	0.2805	0.3008	0.6503	0.3166	0	0	0
Standard Errors	0.0672	0.0757	0.0200	0.0846	0.1150	0.1978	0.1616	0.9265	0.5419	0	0	0
p values	0.0047	0.0001	0.0000	0.1319	0.0419	0.0781	0.0313	0.2404	0.2795	0	0	0
Beta 4	0	0	0	**0.2410**	**0.3958**	**0.9530**	0	0	0	0	0	0
Standard Errors	0	0	0	0.0557	0.1882	0.0918	0	0	0	0	0	0
p values	0	0	0	0.0000	0.0181	0.0000	0	0	0	0	0	0
RQ	108.54	120.88	140.18	106.78	119.73	140.40	107.77	121.55	140.26	120.71	135.93	150.87
Hits in-sample (%)	0.9950	0.9950	1.0305	0.9950	1.0305	1.0305	0.9594	0.9950	0.9950	0.9239	0.9950	0.9594
Hits out-of-sample (%)	0.4000	0.4000	1.7999	0.4000	0.8000	0.6000	0.4000	0.4000	1.6000	1.4000	1.4000	1.6000
DQ in-sample (p values)	0.0436	0.0411	0.0001	0.0337	0.4633	0.0969	0.3498	0.6532	0.0083	0.0064	0.0218	0.0000
DQ out-of-sample (p values)	0.9272	0.9234	0.0000	0.9354	0.9851	0.9895	0.9204	0.9252	0.0564	0.2279	0.0001	0.0000

	CAC40	DAX30	ASE	CAC40	DAX30	ASE	CAC40	DAX30	ASE	CAC40	DAX30	ASE
Beta 1	**0.0187**	**0.0275**	**0.0182**	**0.0250**	0.0364	**0.2374**	0.0241	0.0353	0.0458	**0.6445**	**0.5753**	**0.7951**
Standard Errors	0.01466	0.0155	0.0902	0.00075	0.0164	0.0802	0.0281	0.0303	0.0220	0.0559	0.0651	0.0779
p values	0.1007	0.0388	0.0213	0.0000	0.0134	0.0015	0.1947	0.1218	0.0191	0.0000	0.0000	0.0000
Beta 2	**0.9055**	**0.9177**	**0.9242**	**0.9281**	**0.9300**	**0.6963**	**0.9090**	**0.9129**	**0.8834**	0	0	0
Standard Errors	0.0176	0.0214	0.0116	0.02159	0.0201	0.0445	0.0113	0.0076	0.0086	0	0	0
p values	0.0000	0.0000	0.0000	0.0000	0.0000	0.0000	0.0000	0.0000	0.0000	0	0	0
Beta 3	**0.1819**	**0.1515**	**0.1516**	0.0458	**0.0281**	0.1561	0.2462	0.2361	0.3126	0	0	0
Standard Errors	0.0259	0.0362	0.0217	0.0542	0.0366	0.0530	0.1774	0.4049	0.1016	0	0	0
p values	0.0000	0.0000	0.0000	0.1988	0.2211	0.0163	0.0862	0.2798	0.0104	0	0	0
Beta 4	0	0	0	**0.2111**	**0.2066**	**0.6947**	0	0	0	0	0	0
Standard Errors	0	0	0	0.0470	0.0487	0.0510	0	0	0	0	0	0
p values	0	0	0	0.0000	0.0000	0.0000	0	0	0	0	0	0
RQ	403.02	442.77	477.18	397.67	434.41	460.61	401.72	442.04	475.46	412.62	454.35	486.40
Hits in-sample (%)	5.0103	5.0104	5.0106	5.0106	5.0106	4.9751	5.0106	5.0106	5.0461	4.6197	4.8685	4.7974
Hits out-of-sample (%)	5.4000	4.4000	4.0000	5.8000	5.2000	4.4000	5.4000	4.2000	4.0000	5.4000	5.0000	4.8000
DQ in-sample (p values)	0.8046	0.0709	0.0000	0.8864	0.2294	0.1515	0.7047	0.0405	0.0000	0.6884	0.9833	0.0036
DQ out-of-sample (p values)	0.8359	0.7356	0.1345	0.9849	0.2010	0.7573	0.8217	0.6531	0.1208	0.8451	0.4071	0.0167

Note: Significant coefficients at 5% level of significance are given in bold; shaded boxes denote rejection from the DQ test at 1% significance level.

Table 2. Estimates and Relevant Statistics for the four Conditional Autoregressive Value at Risk Models-Pre Crisis period (03/01/1995-14/9/2007).

1% VaR	Symmetric Absolute Value			Asymmetric slope			Indirect GARCH			Improved Asymmetric		
	FTSE100	S & P500	NIKKEI	FTSE100	S & P500	NIKKEI	FTSE100	S & P500	NIKKEI	FTSE100	S & P5000	NIKKEI
Beta 1	**0.0771**	**0.2116**	**0.0417**	0.0571	**0.1142**	**0.1190**	0.1271	0.5825	0.0781	**0.8313**	**0.0029**	**0.8032**
Standard Errors	0.0317	0.1274	0.0225	0.0192	0.0546	0.0344	0,0923	0.4197	0.0362	0.1505	0.1102	0.1937
p values	0.0074	0.0484	0.0323	0.0015	0.0183	0.0002	0.0844	0.0825	0.0154	0.0000	0.4942	0.0006
Beta 2	**0.8774**	**0.8467**	**0.9419**	**0.9162**	**0.8923**	**0.9148**	**0.8728**	**0.8662**	**0.9424**	0	0	0
Standard Errors	0.0296	0.0543	0.0202	0.0269	0.0305	0.0353	0.0149	0.0431	0.0681	0	0	0
p values	0.0000	0.0000	0.0000	0.0000	0.0000	0.0000	0.0000	0.0000	0.0000	0	0	0
Beta 3	**0.2995**	**0.2979**	**0.1357**	**0.0735**	**0.1649**	**–0.0829**	0.6919	0.5298	0.2738	0	0	0
Standard Errors	0.1027	0.0877	0.0500	0.1195	0.0509	0.0522	1.4580	0.1836	0.3740	0	0	0
p values	0.0017	0.0004	0.0033	0.2691	0.0006	0.0563	0.3175	0.0019	0.2320	0	0	0
Beta 4	0	0	0	**0.3049**	**0.3321**	**0.3333**	0	0	0	0	0	0
Standard Errors	0	0	0	0.0601	0.0889	0.1949	0	0	0	0	0	0
p values	0	0	0	0.0000	0.0436	0.0000	0	0	0	0	0	0
RQ	84.68	119.32	97.44	82.53	116.98	94.39	84.82	119.10	97.06	92.19	134.12	100.70
Hits in-sample(%)	0.9502	0.9502	0.9503	1.0305	1.0305	1.0305	1.0305	0.9509	0.9502	0.9503	0.7871	0.9594
Hits out-of-sample(%)	1.4000	1.0000	1.2000	1.4000	1.0000	0.8000	1.6000	1.0000	1.2000	1.0000	0.9000	1.2000
DQ in-sample (p values)	0.9256	0.2271	0.5227	0.8979	0.9173	0.0398	0.9073	0.6495	0.6423	0.3766	0.5360	0.6310
DQ out-of-sample (p values)	0.4816	0.5094	0.9955	0.2668	0.4067	0.9853	0.7568	0.4826	0.1535	0.9652	0.8716	0.8726
5%VaR	**Symmetric Absolute Value**			**Asymmetric slope**			**Indirect GARCH**			**Improved Asymmetric**		

	FTSE100	S & P500	NIKKEI	FTSE100	S & P500	NIKKEI	FTSE100	S & P500	NIKKEI	FTSE100	S & P500	NIKKEI
Beta 1	**0.0083**	**0.0399**	**0.0108**	**0.0245**	**0.0417**	0.0303	**0.0183**	**0.1439**	0.0102	**0.0655**	**0.2736**	**0.3166**
Standard Errors	0.0142	0.0312	0.0076	0.0092	0.0025	0.0074	0.0176	0.0896	0.0080	0.0501	0.0564	0.0433
p values	0.2770	0.1005	0.0783	0.0004	0.0500	0.0000	0.1483	0.0054	0.1011	0.0000	0.0000	0.0000
Beta 2	**0.9157**	**0.9278**	**0.0965**	**0.9211**	**0.9351**	**0.9593**	**0.9023**	**0.8972**	**0.9504**	0	0	0
Standard Errors	0.0200	0.0205	0.0082	0.0141	0.0267	0.0094	0.0110	0.0164	0.0039	0	0	0
p values	0.0000	0.0000	0.0000	0.0000	0.0000	0.0000	0.0000	0.0000	0.0000	0	0	0
Beta 3	**0.1753**	**0.1225**	**0.0596**	**0.0319**	**0.0438**	**−0.1942**	0.2719	**0.2316**	0.0755	0	0	0
Standard Errors	0.0282	0.0382	0.0109	0.0337	0.0485	0.0185	0.1314	0.2068	0.1467	0	0	0
p values	0.0000	0.0000	0.0000	0.1719	0.1832	0.1473	0.1926	0.1313	0.3032	0	0	0
Beta 4	0	0	0	**0.2616**	**0.1651**	**0.1193**	0	0	0	0	0	0
Standard Errors	0	0	0	0.0364	0.0736	0.0209	0	0	0	0	0	0
p values	0	0	0	0.0000	0.0124	0.0000	0	0	0	0	0	0
RQ	314.29	427.85	325.08	308.06	424.61	316.97	312.69	427.65	326.09	317.00	434.12	322.50
Hits in-sample (%)	5.0103	5.0105	4.9751	5.0103	4.9751	5.0102	5.0102	5.0102	5.0106	4.7619	5.0023	4.9951
Hits out-of-sample (%)	5.2000	5.0000	5.4000	5.6000	5.2000	4.0000	5.2000	4.6000	5.0000	5.2000	5.0000	5.2000
DQ in-sample (p values)	0.0563	0.6356	0.0004	0.8472	0.7184	0.0085	0.2569	0.8704	0.0089	0.1741	0.0096	0.0031
DQ out-of-sample (p values)	0.7984	0.0052	0.0523	0.9607	0.0550	0.0217	0.7945	0.0079	0.6462	0.5476	0.0039	0.7869

Note: Significant coefficients at 5% level of significance are given in bold; shaded boxes denote rejection from the DQ test at 1% significance level.

Table 3. Estimates and Relevant Statistics for the four Conditional Autoregressive Value at Risk Models-Crisis period (17/09/2007-31/08/2009).

1% VaR	Symmetric Absolute Value			Asymmetric slope			Indirect GARCH			Improved Asymmetric		
	CAC40	DAX30	ASE	CAC40	DAX30	ASE	CAC40	DAX30	ASE	CAC40	DAX30	ASE
Beta 1	**0.6502**	**0.5381**	**0.4167**	0.3279	**0.1942**	**0.2485**	0.3999	3.481	1.8874	**0.6788**	**0.7119**	**0.5981**
Standard Errors	0.3348	0.2744	0.4141	0.1609	0.1355	0.0822	0.4188	1.3093	0.9342	0.4077	0.3300	0.2623
p values	0.0266	0.0249	0.1570	0.0208	0.0758	0.0012	0.1698	0.0039	0.0216	0.0523	0.4999	0.0122
Beta 2	**0.7646**	**0.8191**	**0.8575**	**0.7854**	**0.8503**	**0.8224**	**0.9198**	**0.7856**	**0.8181**	0	0	0
Standard Errors	0.1136	0.0927	0.1314	0.1685	0.0961	0.0656	0.0317	0.0417	0.0580	0	0	0
p values	0.0000	0.0000	0.0000	0.0000	0.0000	0.0000	0.0000	0.0000	0.0000	0	0	0
Beta 3	**0.3632**	**0.2549**	**0.2730**	0.2388	**0.1933**	0.1297	0.3270	0.5141	0.7821	0	0	0
Standard Errors	0.2234	0.1612	0.2324	0.4098	0.3126	0.1840	0.9344	1.3942	1.7329	0	0	0
p values	0.0052	0.0568	0.1200	0.2800	0.2683	0.2404	0.3631	0.3562	0.3258	0	0	0
Beta 4	0	0	0	**0.6574**	**0.4570**	**0.6590**	0	0	0	0	0	0
Standard Errors	0	0	0	0.8213	0.1996	0.2801	0	0	0	0	0	0
p values	0	0	0	0.2116	0.0110	0.0093	0	0	0	0	0	0
RQ	261.58	254.60	241.20	256.42	249.58	224.76	264.52	261.83	239.45	318.20	295.96	308.57
Hits in-sample (%)	0.9732	1.2165	0.9732	1.2162	0.7299	0.9723	0.9732	1.2165	0.9732	0.9912	1.0022	0.7299
Hits out-of-sample (%)	0.0000	0.0000	0.0000	1.4000	2.000	2.0000	0.0000	0.6000	0.0000	1.0000	0.9000	1.2000
DQ in-sample (p values)	0.9974	0.0355	0.9977	0.9908	0.9972	0.9979	0.1113	0.9975	0.0005	0.9995	0.9823	0.9783
DQ out-of-sample (p values)	0.0000	1.0000	0.0000	0.0372	0.1133	0.0102	1.0000	1.0000	0.9884	0.9987	0.9286	0.9887
5%VaR	**Symmetric Absolute Value**			**Asymmetric slope**			**Indirect GARCH**			**Improved Asymmetric**		

	CAC40	DAX30	ASE	CAC40	DAX30	ASE	CAC40	DAX30	ASE	CAC40	DAX30	ASE
Beta 1	**0.0941**	**0.0962**	**0.1872**	**0.0719**	0.0809	**0.1241**	0.3466	0.3161	0.5474	**1.0359**	**0.9205**	**1.2622**
Standard Errors	0.0822	0.0763	0.0972	0.0811	0.0505	0.06613	0.2743	0.2200	0.3831	0.1394	0.2182	0.1128
p values	0.1263	0.1037	0.0300	0.1877	0.0546	0.0302	0.1032	0.0754	0.0765	0.0000	0.0000	0.0000
Beta 2	**0.8604**	**0.8729**	**0.7858**	**0.9070**	**0.9172**	**0.8449**	**0.8753**	**0.8361**	**0.7890**	0	0	0
Standard Errors	0.0632	0.1033	0.0946	0.0608	0.0473	0.0693	0.0285	0.0346	0.0292	0	0	0
p values	0.0000	0.0000	0.0000	0.0000	0.0000	0.0000	0.0000	0.0000	0.0000	0	0	0
Beta 3	**0.2741**	**0.2379**	**0.4056**	0.0871	**−0.0199**	0.1277	0.3031	0.4524	0.6012	0	0	0
Standard Errors	0.1386	0.2354	0.2383	0.1830	0.1401	0.1435	1.1754	0.7757	1.2347	0	0	0
p values	0.0239	0.1561	0.0443	0.3170	0.4433	0.1868	0.3982	0.2798	0.3131	0	0	0
Beta 4	0	0	0	**0.2511**	**0.2763**	**0.4124**	0	0	0	0	0	0
Standard Errors	0	0	0	0.0747	0.0753	0.1576	0	0	0	0	0	0
p values	0	0	0	0.0000	0.0000	0.0000	0	0	0	0	0	0
RQ	91.00	88.65	89.61	89.58	85.70	87.16	90.08	87.35	89.46	94.63	92.81	97.11
Hits in-sample (%)	4.8661	4.9921	5.1009	5.0000	4.8861	4.9721	5.0332	4.9443	5.0332	5.0028	5.3527	4.8996
Hits out-of-sample (%)	3.4000	4.0000	4.0000	7.4000	7.0000	4.0000	5.0000	4.0000	4.0000	5.4000	5.2000	4.9000
DQ in-sample (p values)	0.2065	0.1812	0.1548	0.4886	0.5083	0.4914	0.3417	0.1441	0.5200	0.1248	0.3272	0.9379
DQ out-of-sample (p values)	0.8921	0.7963	0.0085	0.9610	0.4468	0.1733	0.9908	0.7236	0.1644	0.8056	0.90049	0.1987

Note: Significant coefficients at 5% level of significance are given in bold; shaded boxes denote rejection from the DQ test at 1% significance level.

Table 4. Estimates and Relevant Statistics for the four Conditional Autoregressive Value at Risk Models-Crisis period (17/09/2007-31/08/2009).

1% VaR	**Symmetric Absolute Value**			**Asymmetric slope**			**Indirect GARCH**			**Improved Asymmetric**		
	FTSE100	**S & P500**	**NIKKEI**	**FTSE100**	**S & P500**	**NIKKEI**	**FTSE100**	**S & P500**	**NIKKEI**	**FTSE100**	**S & P500**	**NIKKEI**
Beta 1	**0.2868**	**0.7739**	**0.1298**	0.3517	**0.6669**	**0.1312**	0.2294	0.2393	0.4768	**0.8891**	**0.8624**	**−0.4942**
Standard Errors	0.3114	0.3811	0.0922	0.3444	0.3788	0.09592	0.6142	1.5304.	0.3544	0.1665	0.5814	0.0889
p values	0.1785	0.0211	0.0795	0.1535	0.0391	0.0855	0.3543	0.0588	0.0892	0.0000	0.0000	0.0000
Beta 2	**0.8244**	**0.7134**	**0.9147**	**0.8751**	**0.7470**	**0.9134**	**0.9229**	**0.7833**	**0.9192**	0	0	0
Standard Errors	0.1594	0.1088	0.0346	0.1192	0.1197	0.0408	0.0386	0.0456	0.0191	0	0	0
p values	0.0000	0.0000	0.0000	0.0000	0.0000	0.0000	0.0000	0.0000	0.0000	0	0	0
Beta 3	**0.4664**	**0.6023**	**0.1950**	**−0.0553**	**0.3265**	**0.2025**	0.5525	0.9251	0.3574	0	0	0
Standard Errors	0.4665	0.2439	0.0621	0.1509	0.3387	0.1795	1.5523	2.4195	3.0159	0	0	0
p values	0.1583	0.0067	0.0008	0.3570	0.1674	0.1295	0.3609	0.3511	0.4582	0	0	0
Beta 4	0	0	0	**0.3918**	**0.7023**	**0.1939**	0	0	0	0	0	0
Standard Errors	0	0	0	0.2489	0.2802	0.0649	0	0	0	0	0	0
p values	0	0	0	0.0577	0.0061	0.0014	0	0	0	0	0	0
RQ	25.33	27.88	25.42	23.41	26.25	25.42	25.46	27.14	25.33	31.67	42.85	35.71
Hits in-sample (%)	0.9732	0.9732	0.9732	0.9732	1.2165.	0.9732	0.9732	0.7299	0.7299	0.7299	0.7299	0.8661
Hits out-of-sample (%)	0.0000	0.0000	1.2000	0.0000	0.0000	0.0000	0.0000	0.0000	0.0000	0.0000	0.0000	0.0000
DQ in-sample (p values)	0.9972	0.0004	0.9972	0.9966	0.9910	0.9974	0.9980	0.9972	0.9982	0.3550	0.9554	0.9999
DQ out-of-sample (p values)	0.0000	0.0000	0.0000	1.0000	1.0000	1.0000	0.0000	0.0000	0.0000	0.9601	0.9073	0.7836
5%VaR	**Symmetric Absolute Value**			**Asymmetric slope**			**Indirect GARCH**			**Improved Asymmetric**		

	FTSE100	S & P500	NIKKEI	FTSE100	S & P500	NIKKEI	FTSE100	S & P500	NIKKEI	FTSE100	S & P500	NIKKEI
Beta 1	**0.1237**	**0.1161**	**0.2030**	**0.0751**	**0.0597**	0.1155	**0.3908**	**0.3288**	0.6780	**0.7424**	**0.1201**	**0.9932**
Standard Errors	0.0990	0.1005	0.1660	0.0864	0.0433	0.0851	0.2890	0.5494	0.4233	0.2206	0.1360	0.1719
p values	0.1057	0.1240	0.1107	0.1925	0.0840	0.0874	0.0881	0.2747	0.0546	0.0000	0.0000	0.0000
Beta 2	**0.8807**	**0.8537**	**0.8705**	**0.9218**	**0.9401**	**0.9276**	**0.8785**	**0.8471**	**0.8451**	0	0	0
Standard Errors	0.0624	0.0929	0.0911	0.0664	0.0325	0.0472	0.0382	0.0266	0.0556	0	0	0
p values	0.0000	0.0000	0.0000	0.0000	0.0000	0.0000	0.0000	0.0000	0.0000	0	0	0
Beta 3	**0.1928**	**0.2957**	**0.1840**	**0.0205**	**−0.0852**	**−0.0172**	0.2485	**0.5331**	0.3501	0	0	0
Standard Errors	0.0934	0.2615	0.1284	0.1709	0.0716	0.0964	0.4282	1.8901	1.0478	0	0	0
p values	0.0195	0.1290	0.0758	0.4522	0.1169	0.4289	0.2808	0.3889	0.3691	0	0	0
Beta 4	0	0	0	**0.2318**	**0.2698**	**0.1878**	0	0	0	0	0	0
Standard Errors	0	0	0	0.1075	0.1063	0.0730	0	0	0	0	0	0
p values	0	0	0	0.0155	0.0056	0.0051	0	0	0	0	0	0
RQ	88.71	109.58	97.62	87.66	104.28	95.88	88.52	104.97	96.27	91.53	108.34	95.39
Hits in-sample (%)	4.6228	5.1094	4.8661	4.6228	4.3795	5.1094	4.8661	5.1094	5.3527	5.1094	5.3525	5.1094
Hits out-of-sample (%)	3.0000	2.0000	3.0000	4.0000	7.0000	4.0000	3.0000	3.0000	3.0000	2.0000	3.0000	3.0000
DQ in-sample (p values)	0.5261	0.0786	0.3170	0.6396	0.6416	0.2993	0.3909	0.0936	0.8336	0.6592	0.0923	0.5858
DQ out-of-sample (p values)	0.8829	0.8388	0.9855	0.9798	0.1688	0.9907	0.9599	0.7738	0.9842	0.9336	0.7797	0.9630

Note: Significant coefficients at 5% level of significance are given in bold; shaded boxes denote rejection from the DQ test at 1% significance level.

Table 5. Estimates and Relevant Statistics for the four Conditional Autoregressive Value at Risk Models-Post Crisis period (01/09/2009-23/08/2013).

1% VaR	Symmetric Absolute Value			Asymmetric slope			Indirect GARCH			Improved Asymmetric		
	CAC40	DAX30	ASE	CAC40	DAX30	ASE	CAC40	DAX30	ASE	CAC40	DAX30	ASE
Beta 1	**0.1557**	**0.1507**	**−0.0346**	0.1229	**0.1649**	**−0.0346**	0.7035	0.3601	0.4845	**1.1375**	**1.0992**	**−0.1855**
Standard Errors	0.0819	0.0755	0.0299	0.0649	0.0844	0.0293	0.3130	0.2535	0.7815	0.1839	0.2230	0.0862
p values	0.0286	0.0639	0.1238	0.0227	0.0253	0.1187	0.0123	0.0777	0.2676	0.0000	0.0000	0.0156
Beta 2	**0.8440**	**0.8580**	**0.9870**	**0.8440**	**0.8393**	**0.9866**	**0.8008**	**0.8422**	**0.9647**	0	0	0
Standard Errors	0.0488	0.0456	0.0075	0.0445	0.0699	0.0076	0.0304	0.0419	0.0232	0	0	0
p values	0.0000	0.0000	0.0000	0.0000	0.0000	0.0000	0.0000	0.0000	0.0000	0	0	0
Beta 3	**0.3532**	**0.3372**	**0.0660**	0.2015	**0.1594**	0.0000	0.0000	0.0000	0.0000	0	0	0
Standard Errors	0.0963	0.0714	0.0213	0.2624	0.1594	0.0368	0.5401	0.7061	03636	0	0	0
p values	0.0001	0.0000	0.0000	0.0221	0.0834	0.0206	0.0512	0.1360	0.0333	0	0	0
Beta 4	0	0	0	**0.5125**	**0.5614**	**0.0603**	0	0	0	0	0	0
Standard Errors	0	0	0	0.2167	0.2765	0.0222	0	0	0	0	0	0
p values	0	0	0	0.0090	0.0211	0.0031	0	0	0	0	0	0
RQ	32.52	30.83	53.73	32.07	29.43	53.69	33.08	31.34	54.46	36.66		194.03
Hits in-sample (%)	0.9535	1.0727	1.0727	0.9535	0.9535	0.9535	0.9535	1.1918	0.9535	0.8343	1.1918	0.7151
Hits out-of-sample (%)	1.5000	1.5000	1.0000	2.0000	1.5000	1.0000	1.0000	1.5000	0.5000	1.0000	1.0000	0.0000
DQ in-sample (p values)	0.9927	0.9812	0.9852	0.9912	0.9960	0.9923	0.9931	0.9777	0.9939	0.0000	0.0000	0.9992
DQ out-of-sample (p values)	0.8752	0.9369	0.9996	0.2383	0.9171	0.9987	0.9961	0.7984	0.9245	0.4323	0.3978	1.0000
5%VaR	**Symmetric Absolute Value**			**Asymmetric slope**			**Indirect GARCH**			**Improved Asymmetric**		

	CAC40	DAX30	ASE	CAC40	DAX30	ASE	CAC40	DAX30	ASE	CAC40	DAX30	ASE
Beta 1	**0.0903**	**0.1515**	**0.1956**	**0.0712**	0.1352	**0.2172**	0.4273	0.4060	0.3369	**1.0679**	**0.9270**	**0.5362**
Standard Errors	0.0754	0.0768	0.1250	0.0450	0.0511	0.1119	0.3500	0.2322	0.4501	0.1367	0.1063	0.1202
p values	0.1155	0.0243	0.0589	0.0566	0.0041	0.0261	0.1110	0.0402	0.0227	0.0000	0.0000	0.0000
Beta 2	**0.8218**	**0.7964**	**0.8538**	**0.8939**	**0.8575**	**0.8543**	**0.7197**	**0.7711**	**0.8884**	0	0	0
Standard Errors	0.0688	0.0683	0.0543	0.0421	0.0532	0.0429	0.0688	0.0505	0.0440	0	0	0
p values	0.0000	0.0000	0.0000	0.0000	0.0000	0.0000	0.0000	0.0000	0.0000	0	0	0
Beta 3	**0.3425**	**0.3412**	**0.2004**	0.0124	**0.0640**	0.1334	0.7669	0.5404	0.2314	0	0	0
Standard Errors	0.1696	0.1332	0.0822	0.1040	0.1511	0.0548	0.6963	1.4738	0.1234	0	0	0
p values	0.0217	0.0052	0.0074	0.4522	0.3357	0.0075	0.1353	0.3569	0.0303	0	0	0
Beta 4	0	0	0	**0.3351**	**0.3450**	**0.2095**	0	0	0	0	0	0
Standard Errors	0	0	0	0.0772	0.0910	0.0629	0	0	0	0	0	0
p values	0	0	0	0.0000	0.0000	0.0000	0	0	0	0	0	0
RQ	132.85	125.47	196.17	128.13	121.83	195.93	134.51	126.91	195.65	140.38	132.21	204.10
Hits in-sample (%)	5.0059	5.2443	5.0059	4.8867	5.1251	5.0059	5.0059	5.0059	5.0059	4.6483	4.8867	4.7675
Hits out-of-sample (%)	3.5000	2.5000	6.5000	4.0000	3.5000	7.0000	3.0000	2.5000	6.5000	4.5000	4.5000	5.0000
DQ in-sample (p values)	0.9101	0.4057	0.4940	0.4034	0.7877	0.7199	0.9703	0.6512	0.8859	0.3060	0.5515	0.2326
DQ out-of-sample (p values)	0.7546	0.8325	0.0307	0.6341	0.8023	0.0956	0.8411	0.8224	0.0799	0.3787	0.5855	0.2147

Note: Significant coefficients at 5% level of significance are given in bold; shaded boxes denote rejection from the DQ test at 1% significance level.

Table 6. Estimates and Relevant Statistics for the four Conditional Autoregressive Value at Risk Models-Post Crisis period (01/09/2009-23/08/2013).

1%VaR	Symmetric Absolute Value			Asymmetric slope			Indirect GARCH			Improved Asymmetric		
	FTSE100	S & P500	NIKKEI	FTSE100	S & P500	NIKKEI	FTSE100	S & P500	NIKKEI	FTSE100	S & 500	NIKKEI
Beta 1	**0.1054**	**0.7418**	**0.1507**	0.0590	**0.2325**	**0.2454**	0.3603	4.6942	0.5913	**1.3272**	**0.0000**	**0.9730**
Standard Errors	0.0646	0.4360	0.0813	0.0470	0.0701	0.1123	0.1982	2.5886	0.4690	0.2890	0.3152	0.1493
p values	0.0513	00444	0.0319	0.1049	0.0000	0.0144	0.0345	0.0348	0.1037	0.0000	0.5000	0.0000
Beta 2	**0.8083**	**0.4668**	**0.8883**	**0.9090**	**0.8172**	**0.8576**	**0.7551**	**0.2913**	**0.7389**	0	0	0
Standard Errors	0.7913	0.1466	0.0536	0.0627	0.0363	0.0703	0.0354	0.2387	0.0495	0	0	0
p values	0.0000	0.0000	0.0000	0.0000	0.0000	0.0000	0.0000	0.1111	0.0000	0	0	0
Beta 3	**0.5129**	**1.2763**	**0.2207**	**0.0737**	**−0.1721**	**−0.1511**	1.3745	2.8171	1.7650	0	0	0
Standard Errors	0.2492	0.1433	0.1168	0.1936	0.1218	0.1177	1.2085	0.4882	7.1408	0	0	0
p values	0.0197	0.0000	0.0293	0.3515	0.0788	0.0996	0.1276	0.0000	0.4023	0	0	0
Beta 4	0	0	0	**0.3672**	**0.9914**	**0.6778**	0	0	0	0	0	0
Standard Errors	0	0	0	0.1639	0.0776	0.2708	0	0	0	0	0	0
p values	0	0	0	0.0125	0.0000	0.0061	0	0	0	0	0	0
RQ	24.92	35.21	30.17	24.40	31.83	28.16	25.71	32.89	29.21	28.23	39.98	34.49
Hits in-sample (%)	1.0727	1.0727	1.0727	0.9535	1.0727	0.9535	0.9535	0.9535	0.9535	0.8343	0.5959	0.9535
Hits out-of-sample (%)	1.0000	1.5000	0.0000	2.5000	5.0000	0.5000	1.0000	2.0000	0.5000	1.5000	3.5000	1.0000
DQ in-sample (p values)	0.9852	0.0444	0.0431	0.9925	0.0446	0.0299	0.0168	0.9918	0.0244	0.0088	0.0003	0.0000
DQ out-of-sample (p values)	0.9998	0.9858	1.0000	0.0000	0.0000	0.9963	0.9975	0.8029	0.9975	0.8150	0.0000	0.7612
5%VaR	**Symmetric Absolute Value**			**Asymmetric slope**			**Indirect GARCH**			**Improved Asymmetric**		

	FTSE100	S & P500	NIKKEI	FTSE100	S & P500	NIKKEI	FTSE100	S & P500	NIKKEI	FTSE100	S & P500	NIKKEI
Beta 1	**0.1235**	**0.1362**	**0.0620**	**0.0653**	**0.1513**	0.0792	**0.2682**	**1.9670**	0.1131	**0.7709**	**0.5021**	**0.9710**
Standard Errors	0.0618	0.1456	0.0442	0.0391	0.0718	0.0294	0.1207	1.7496	0.0721	0.0900	0.1040	0.0620
p values	0.0229	0.1748	0.0804	0.0477	0.0175	0.0035	0.0131	0.1304	0.0583	0.0000	0.0000	0.0000
Beta 2	**0.8346**	**0.8729**	**0.8718**	**0.9300**	**0.8852**	**0.8749**	**0.8152**	**0.3104**	**0.8462**	0	0	0
Standard Errors	0.0485	0.1003	0.0322	0.0397	0.0647	0.0386	0.0338	0.4300	0.0274	0	0	0
p values	0.0000	0.0000	0.0000	0.0000	0.0000	0.0000	0.0000	0.2351	0.0000	0	0	0
Beta 3	**0.2449**	**0.1255**	**0.2242**	**−0.0836**	**−0.0598**	**0.0109**	0.3423	**0.6348**	0.3750	0	0	0
Standard Errors	0.0711	0.0812	0.0478	0.0778	0.0733	0.1156	0.3180	0.3063	0.4523	0	0	0
p values	0.0002	0.0612	0.0000	0.1411	0.2073	0.4623	0.1408	0.0191	0.2035	0	0	0
Beta 4	0	0	0	**0.2509**	**0.2190**	**0.3856**	0	0	0	0	0	0
Standard Errors	0	0	0	0.1121	0.1323	0.1503	0	0	0	0	0	0
p values	0	0	0	0.0125	0.0489	0.0052	0	0	0	0	0	0
RQ	100.46	117.82	108.60	97.71	114.60	106.80	101.03	116.66	109.28	105.47	118.72	111.07
Hits in-sample (%)	4.8867	5.1251	5.1251	5.2443	5.1251	5.0059	5.0059	5.0059	5.1251	4.8867	4.6483	4.8867
Hits out-of-sample (%)	3.5000	6.5000	3.0000	3.5000	4.8000	3.5000	4.0000	7.5000	3.0000	5.0000	7.0000	4.0000
DQ in-sample (p values)	0.8704	0.5473	0.3859	0.3517	0.3995	0.3023	0.8221	0.5189	0.4050	0.5467	0.5702	0.0752
DQ out-of-sample (p values)	0.7867	0.0961	0.5864	0.7419	0.0089	0.7737	0.8776	0.2265	0.5344	0.3512	0.0000	0.3686

Note: Significant coefficients at 5% level of significance are given in bold; shaded boxes denote rejection from the DQ test at 1% significance level.

We then turn to the results for the 5% VaR. We first discuss the results for the CAC40, the DAX30 and the ASE stock indices. The in-sample hits range from 4.9% to 5.1% for the Symmetric Absolute Value, the Asymmetric Slope and the Indirect GARCH models, while the I-CAViaR specification is on target in most cases as well. The out-of sample forecasts are very accurate for the case of DAX40 when we apply any of these three models, and also for the case of CAC40 when we apply the Indirect GARCH specification. When we turn to the case of the FTSE100, S & P500 and NIKKEI225 indices note that we obtain estimates which are extremely close to the value of 5%, which is taken as evidence that they describe the evolution of the tail for all cases under consideration. Looking into the out-of-sample forecasts, the performance is similar to the one obtained in the first set of market indices. Finally, a similar pattern emerges for all three specifications as well the I-CAViaR model when we examine the five stock indices.

A final comment we make is that in most cases the estimation of the Asymmetric Slope model gives coefficient estimates for the negative lagged returns which are always statistically significant while the estimates associated with the positive returns are not significantly different from zero. Therefore, we may argue there are possible strong asymmetric influences on VaR measures of lagged returns. This finding is further reinforced by the estimates of the I-CAViaR model.

The overall results from the present analysis show that the DQ test statistics select different CAViaR specifications for different confidence internals which may lead to the argument that the process guiding the tail behaviour changes over time.

With respect to the empirical loss function results we observe that the values of the $Q(\theta)$ I-CAViaR model are always less that of the three specifications due to Engle & Managanelli (2004) .

5. SUMMARY AND CONCLUDING REMARKS

The recent financial crisis of 2007-2009 has brought in surface the need for more accurate measurement of the downside risk of financial institutions. A significant contribution to this increased volatility has been the substantial rise of capital flows from mature markets towards the emerging markets of Southeast Asia and the economies of transition. The 1997-1998 financial crisis as well as the bankruptcy of several financial institutions during the 1990s had shown quite convincingly how important the development is for the stability of the global financial system and adoption of the appropriate mechanisms for measuring market risk. However, the recent financial turbulence and the reversal of capital flows in an effort of the hedge funds to liquidate their investment in emerging markets have once again led to increase volatility in the global financial markets.

The Riskemetrics methodology by Morgan Stanley (1996) provided the first econometric methodology for the measurement of market risk based on the concept of VaR which was introduced in Banking Supervision's (1996) "Amendment to the Capital Accord to Incorporate Market Risks". This led to a fast growing literature on the development of alternative methodologies to measure VaR. The present paper utilized the CAViaR modeling procedure, which had been proposed by Engle & Manganelli (2004) . This is a semiparametric method which shifts the analysis of developing a good measure of the VaR from the distribution of the portfolio returns directly to the behavior of the quantile. This methodology is considered to have several advantages over competing VaR methodologies like the GARCH models and the Extreme Value Theory.

We applied this methodology to estimate the VaR using daily observations for the period January 3, 1995 to August 28, 2013. We studied the behaviour of alternative CAViaR specifications for six stock market indices, namely the S & P 500, FTSE 100, DAX30, CAC40, NIKKE225 and the Athens Exchange General Index.

Our overall results led to the conclusion that this methodology provided very accurate measurement of the VaR for all stock price indices. This evidence is based on a battery in-sample and out-of-sample forecasting performance. Furthermore, our findings show that the CAViaR quantile regression models and the I-CAViaR model have shown significant success in predicting the VaR measure for various periods although this performance varies over the three periods before, during and after the 2007-2009 financial crisis. These findings may be of interest to portfolio managers, private and institutional investors as well as hedge funds that are active in mature and emerging stock markets.

ACKNOWLEDGEMENTS

We thank participants at the 6th International Conference on Mathematical and Statistical Methods for Actuarial Sciences and Finance, 22-24 April 2014 and at seminars at Athens University of Economics and Business, European University Institute, Hebrew University of Jerusalem, University of Crete, University of Piraeus and Central Bank of Cyprus for their very helpful comments. Kouretas acknowledges financial support from a Marie CurieTransfer of Knowledge Fellowship of the European Community's Sixth Framework Programme under contract number MTKD-CT-014288, as well as from the Research Committee of the University of Crete under researchgrants #2016, #2030 and #2257. We thank Carol Alexander, Richard Baillie, Manthos Delis, Dimitris Georgoutsos, Dashan Huang, Michael McAleer, Simone Manganelli, Stavros Stavroyiannis, Chris Tsoumas, Elias Tzavalis and Mark Wohar for many helpful comments and discussions. We also thank Dashan Huang and Baimin Yu for providing us with the MATLAB codes for estimating the improved asymmetric CAViaR model. Finally, we thank the editor and an anonymous referee for their constructive comments which improved the manuscript substantially. The usual caveat applies.

REFERENCES

1. Alexander, S. (2005). The Present and Future of Financial Risk Management. Journal of Financial Econometrics, 3, 3-25.
2. Alexander, S., Coleman, T. F., & Li, Y. (2006). Minimizing CVaR and VaR for a Portfolio of Derivatives. Journal of Banking and Finance, 30, 583-605.
3. Aloui, C., & ben Hamida, H. (2014). Modelling and Forecasting Value at Risk and Expected Shortfall for GCC Stock Markets: Do Long Memory, Structural Breaks, Asymmetry, and Fat-Tails Matter? North American Journal of Economics and Finance, 29, 349-380.

4. Angelidis, T., Benos, A., & Degiannakis, S. (2004). The Use of GARCH Models in VaR Estimation. Statistical Methodology, 1, 105-128.

5. Artzner, P., Delbaen, F., Eber, J.-M., & Heath, D. (1997). Thinking Coherently. Risk, 10, 68-71.

6. Artzner, P., Delbaen, F., Eber, J.-M., & Heath, D. (1999). Coherent Measures of Risk. Mathematical Finance, 9, 203-228.

7. Bams, D., Lehnert, T., & Wolff, C. C. P. (2005). An Evaluation Framework for Alternative VaR-Models. Journal of International Money and Finance, 24, 922-958.

8. Bank for International Settlements (1988). International Convergence of Capital Measurement and Capital Standards. BCBS Publication Series, No.4.

9. Banking Supervision's (1996). Amendment to the Capital Accord to Incorporate Market Risks.

10. Bank for International Settlements (1999a). Capital Requirements and Bank Behavior: The Impact of the Basel Accord. BCBS Working Paper Series, No. 1.

11. Bank for International Settlements (1999b). A New Capital Adequacy Framework. BCBS Publications Series, No. 50.

12. Bank for International Settlements (1999c). Supervisory Lesson to Be Drawn from the Asian Crisis. BCBS Working Paper Series, No. 2.

13. Bank for International Settlements (2001). The New Basel Capital Accord. Basel: BIS.

14. Basel Committee on Banking Supervision (1996). Amendment to the Capital Accord to Incorporate Market Risks.

15. Bao, Y., Lee, T., & Saltoglu, B. (2006). Evaluating the Predictive Performance of Value-at-Risk Models in Emerging Markets: A Reality Check. Journal of Forecasting, 25, 101-128.

16. Bekiros, S., & Georgoutsos, D. A. (2005a). Estimation of Value-at-Risk by Extreme Value and Conventional Methods: A Comparative Evaluation of Their Predictive Performance. Journal of International Financial Markets, Institutions and Money, 15, 209-228.

17. Bekiros, S., & Georgoutsos, D. A. (2005b). Extreme Returns and the Contagion Effect between the Foreign Exchange and the Stock Market: Evidence from Cyprus. Multinational Finance Journal, Forthcoming.

18. Boudoukh, J., Richardson, M., & Whitelaw, R. F. (1998). The Best of Both Worlds. Risk, 11, 64-67.

19. Boyle, P., Hardy, M., & Vorst, T. (2005). Life after VaR. Journal of Derivatives, 13, 48-55.

20. Brooks, C., Clare, A. D., Dalle-Mulle, J. W., & Persand, G. (2005). A Comparison of Extreme Value Theory Approaches for Determining Value at Risk. Journal of Empirical Finance, 12, 339-352.

21. Burns, P. (2005). The Quality of Value at Risk via Univariate GARCH. Working Paper, Burns Statistics.

22. Chen, C. W. S., Gerlach, R., Hwang, B. B. K., & McAller, M. (2012). Forecasting Value-at-Risk Using Nonlinear Regression Quantiles and the Intra-Day Range. International Journal of Forecasting, 28, 557-574.

23. Chernozhukov, V. (1999). Specification and Other Tests Processes for Quantile Regression. Stanford University, Mimeograph.

24. Christoffersen, P. F. (1998). Evaluating Interval Forecasts. International Economic Review, 39, 841-864.

25. Danielsson, J., Hartmann, P., & de Vries, C. (1998). The Cost of Conservatism. Risk, 11, 101-103.

26. Danielsson, J., & de Vries, C. (2000). Value-at-Risk and Extreme Returns. Annales d'Economie et de Statistique, 60, 239-270.

27. Drzik, J. (2005). New Directions in Risk Management. Journal of Financial Econometrics, 3, 26-36.

28. Duffie, D., & Pan, J. (1997). An Overview of Value at Risk. Journal of Derivatives, 4, 7-49.

29. Engle, R. F., & Manganelli, S. (2004). CAViaR: Conditional Autoregressive Value at Risk by Regression Quantile. Journal of Business and Economic Statistics, 22, 367-381.

30. Fama, E. F. (1965). The Behaviour of Stock Market Prices. Journal of Business, 38, 34-105.

31. Giacomini, R., & Komunjer, I. (2005). Evaluation and Combination of Conditional Quantile Forecasts. Journal of Economic and Business Statistics, 23, 416-431.

32. Gourieroux, C., & Jasiak, J. (2008). Dynamic Quantile Models. Journal of Econometrics, 147, 198-205.

33. Granger, C. W. J., White, H., & Kamstra, M. (1989). Interval Forecasting: An Analysis Based upon ARCH-Quantiles Estimators. Journal of Econometrics, 40, 87-96.

34. Gerlach, R., Chen, C. W. S., & Chan, N. Y. C. (2011). Bayesian Time-Varying Quantile Forecasting for Value-at-Risk in Financial Markets. Journal of Business and Economic Statistics, 23, 416-431.

35. Guidolin, M., & Timmermann, A. (2006). Term Structure of Risk under Alternative Econometric Specifications. Journal of Econometrics, 131, 285-308.

36. Haas, M., Mittnik, S., & Paolella, M. (2006). Value-at-Risk Prediction: A Comparison of Alternative Strategies. Journal of Financial Econometrics, 4, 53-89.

37. Huang, D., Yu, B., Lu, Z., Fabozzi, F. J., Focardi, S., & Fukushima, M. (2010). Index-Exciting CAViaR: A New Empirical Time-Varying Risk Model. Studies in Nonlinear Dynamics and Econometrics, 14, 1-24.

38. Huang, D., Yu, B., Fabozzi, F. J., & Fukushima, M. (2009). CAViaR-Based Forecast for Oil Price Risk. Energy Economics, 31, 511-518.

39. Hull, J., & White, A. (1998). Value-at-Risk When Daily Changes in Market Variables Are Not Normally Distributed. Journal of Derivatives, 5, 9-19.

40. Jeon, J., & Taylor, J. W. (2013). Using CAViaR Models with Implied Volatility for Value-at-Risk Estimation. Journal of Forecasting, 32, 62-74.

41. Jimenez-Martin, J. A. (2009). The Ten Commandments for Managing Value-at-Risk and the Basel II Accord. Journal of Economic Surveys, 23, 850-855.

42. Jorion, P. (2000). Value at Risk (2nd ed.). New York: McGraw Hill.

43. Koenker, R., & Bassett, G. (1978). Regression Quantiles. Econometrica, 46, 33-50.

44. Koenker, R., & Bassett, G. (1982). Robust Tests for Heteroskedasticity Based on Regression Quantiles. Econometrica, 50, 43-61.

45. Koenker, R., & Zhao, Q. (1996). Conditional Quantile Estimation and Inference for ARCH Models. Econometric Theory, 12, 265-283.

46. Kouretas, G. P., & Zarangas, L. (2005). Conditional Autoregressive Value-at-Risk by Regression Quantiles: Estimating Market Risk for Major Stock Market. Working Paper, Crete: Department of Economics, University of Crete.

47. Kuester, R. F., Mittnik, S., & Paolella, M. (2006). Value-at-Risk Prediction: A Comparison of Alternative Strategies. Journal of Financial Econometrics, 4, 53-89.

48. Longin, F. M. (2000). From VaR to Stress Testing: The Extreme Value Approach. Journal of Banking and Finance, 24, 1097-1130.

49. Longin, F. M. (2001). Beyond the VaR. Journal of Derivatives, 8, 36-48.

50. Mandelbrot, B. (1963). The Variation of Certain Speculative Prices. Journal of Business, 36, 394-419.

51. Manganelli, S., & Engle, R. F. (2004). A Comparison of Value-at-Risk Models in Finance. In G. Szego (Ed.), Risk Measures for the 21st Century (pp. 123-143). Chichester: John Wiley.

52. McAleer, M., & Da Veiga, B. (2008). Forecasting Value-at-Risk with a Parsimonious Portfolio Spillover GARCH (PS- GARCH) Model. Journal of Forecasting, 27, 1-19.

53. McAleer, M., Jimenez-Martin, J. A., & Perez-Amaral, T. (2013). Has the Basel Accord Improved Risk Management during the Global Financial Crisis? North American Journal of Economics and Finance, 26, 250-265.

54. McNeil, A. J., & Frey, R. (2000). Estimation of Tail-Related Risk Measures for Heteroskedastic Financial Time Series: An Extreme Value Approach. Journal of Empirical Finance, 7, 271-300.

55. McNeil, A. J., & Saladin, T. (2000). Developing Scenarios for Future Extreme Losses Using the POT Model. In P. Embrehts (Ed.), Extremes and Integrated Risk Management. London: RISK Publications.

56. Naftci, S. (2000). Value at Risk Calculations, Extreme Events, and Tail Estimation. Journal of Derivatives, 7, 23-37.

57. Pojarlev, M., & Polasek, W. (2000). Value at Risk Estimation for Stock Indices Using the Basle Committee Proposal Form 1995, Mimeograph.

58. Polasek, W., & Pojarlev, M. (2005). VaR Evaluations Based on Volatility Forecasts of GARCH Models, Mimeograph.

59. Portnoy, S. (1991). Asymptotic Behavior of Regression Quantiles in Non-Stationary Dependent Cases. Journal of Multivariate Analysis, 38, 100-113.

60. RiskMetrics (1996). Technical Document. Morgan Guarantee Trust Company of New York.

61. Romero, P. A., Muela, S. B., & Martin, C. L. (2013). A Comprehensive Review of Value at Risk Methodologies. Fundacion De Las Cajas De Ahorros, Documento de Trabajo No 711/2013.

62. Rubia, A., & Sanchis-Marco, L. (2013). On Downside Risk Predictability through Liquidity and Trading Activity: A Dynamic Quantile Approach. International Journal of Forecasting, 29, 202-219.

63. Schaumburg, J. (2012). Predicting Extreme Value at Risk: Nonparametric Quantile Regression with Refinements from Extreme Value Theory. Computational Statistics and Data Analysis, 56, 4081-4096.

64. Schwert, G. W. (1988). Why Does Stock Market Volatility Change over Time? Journal of Finance, 44, 1115-1153.

65. Sener, E., Baronyan, S., & Ali Menguturk, L. (2012). Ranking the Predictive Performances of Value-at-Risk Estimation Methods. International Journal of Forecasting, 28, 849-873.

66. Taylor, S. J. (1986). Modelling Financial Time Series. New York: Wiley and Sons Ltd.

67. White, H. (1994). Estimation, Inference and Specification Analysis. Cambridge: Cambridge University Press.

68. White, H. (2000). A Reality Check for Data Snooping. Econometrica, 68, 1097-1126.

69. Yu, P. L. H., Li, W. K., & Jin, S. (2010). On Some Models for Value-at-Risk. Econometric Reviews, 29, 622-641.

NOTES

[1]For a detailed analysis see the Basel Committee on Banking Supervision's (1996), "Amendment to the Capital Accord to Incorporate Market Risks". Duffie & Pan (1997), Alexander (2005) and Drzik (2005) provide a comprehensive overview of value at risk measures.

[2]See also Bank for International Settlements (1988, 1999a,b,c, 2001). Jimenez-Martin (2009)and McAleer et al. (2013) provide a more recent evaluation of managing value-at-risk under the provisions of the Basle II accord.

[3]CVaR is an alternative risk measure that satisfies the coherency criteria by Artzner et al. (1997, 1999) . Its advantage over VaR measures is that it focuses on both the frequency and the size of extreme events.

[4] Chernozhukov (1999) has derived independently the same dynamic quantile test.

[5] Jorion (2000) and Kuester et al. (2006) provide a complete analysis of the VaR methodology and alternative estimation methodologies. Romero et al. (2013) gives an updated survey of alternative VaR methodologies.

[6]For simplicity we have eliminated the subscript θ from the vector of unknown parameters.

[7]See also Taylor (1986) , Schwert (1988) and Engle (2002).

[8] $(x)^{+} = \max(x,0),\ (x)^{-} = -\min(x,0)$

[9]It is worth noting that the CAViaR specifications are more general than the fitted GARCH models. They can allow for a wide range of assumptions with respect to the error distribution and they can also handle distributions with non-i.i.d. errors.

[10]Following the seminal paper by Koenker & Bassett (1978) , a number of alternative regression quantile models have been developed over the last twenty years that take into account alternative assumptions about the errors. Among others, Koenker & Bassett (1982) allow for the case of heteroskedastic errors, whereas Portnoy (1991) considers the case of non-stationary dependent errors. Furthermore, we have extensions that cover the cases of time series models, simultaneous equations and censored regression models and recently we also have extensions that deal with the case of autoregressive quantiles (Koenker & Zhao, 1996). See Engle & Manganelli (2004) for the relevant literature. All these models differ from the Engle & Manganelli (2004) CAViaR models since they are linear in the parameters.

[11]The complete derivation of the two tests is given in Engle & Manganelli (2004: pp. 370-371) .Granger et al. (1989) and Christoffersen (1998) are among other studies which have developed test statistics for the validity of the forecast model, but as Engle & Manganell (2004) point out, they have low power against misspecification introduced by the presence of serial correlation in the conditional probabilities leading to a quantile measurement error.

[12] Aloui et al. (2014) provide further analysis of forecasting measures whereas Sener et al. (2012)analyze the ranking of predictive performance of value-at-risk estimations methods.

[13]To estimate the Adaptive model we set G = 10, in all cases where G entered the definition of the Adaptive model in section 3.

[14]All assumptions which these models need to satisfy are given in Appendix A of Engle & Manganelli (2004) ; and they refer to asymptotic results but are difficult to verify in finite samples.

[15]As Engle & Manganelli (2004) point out, the lagged hit variables contain the indicator function. Given that the indicator function is Lipschitz continuous it satisfies condition, DQ3 of theorem 4 (Engle & Manganelli, 2004).

[16]The calculation of the two statistics is described in Engle and Manganelli, theorems 3 and 4. Furthermore, for the optimization procedures we adopt the strategy explained in Engle & Manganelli (2004) . The computations were made in Matlab 6.1 using the functions fminsearch and fminunc as the optimization algorithms while the loops to compute the recursive quantile functions were coded in C and have been developed by Manganelli.

6

CHAPTER

A Skewness-Adjusted Binomial Model for Pricing Futures Options—The Importance of the Mean and Carrying-Cost Parameters

Stafford Johnson[1], Amit Sen[2], Brian Balyeat[1]

[1]*Department of Finance, Xavier University, Cincinnati, USA*

[2]*Department of Economics, Xavier University, Cincinnati, USA*

ABSTRACT

In this paper, we extend the Johnson, Pawlukiwicz, and Mehta [1] skewness-adjusted binomial model to the pricing of futures options and examine in some detail the asymptotic properties of the skewness model as it applies to futures and spot options.The resulting skewness-adjusted futures options model shows that for a large number of subperiods, the price of futures options depends not only on the volatility and mean but also on the risk-free rate, asset-yield, and other carrying-cost parameters when skewness exists.

Keywords: Options Pricing; Black-Scholes; Skewness

1. INTRODUCTION

One of the interesting, as well as subtle, features of the Black-Scholes (B-S) [2] model and the binomial option pricing model (BOPM) with a large number of subperiods (n) is that the models depend only on the variance. In these models, the mean is not important in determining the value of spot options and the mean and net carry cost are not important for futures options. These implications of the model, however, depend on the assumption that the logarithmic return of the underlying security is normally distributed. Studies by Johnson, Zuber, and Gandar [3] and [4] have shown that in periods of increasing stock prices or rates, the logarithmic return of stock indexes and interest rates are often characterized by a positive mean and significant negative skewness, and in periods of decreasing prices or rates, the logarithmic returns are often characterized by a negative mean and significant positive skewness. Moreover, several earlier empirical studies have reported that the B-S model consistently underprices options in the presence of skewness; see for instance, Stein and Stein [5], Wiggins [6], and Heston [7].

Jarrow and Rudd [8] and Corrado and Tie Su [9] have extended the B-S model to account for cases in which there is skewness in the underlying security's return distribution. Similarly, Câmara and Chung [10] and Johnson, Pawlukiewicz, and Mehta (JPM) [1] have extended the Cox, Ross, and Rubinstein (CRR) [11] and Rendleman and Bartter (RB) [12] binomial option pricing model to include skewness. In their paper, JPM also show that skewness changes the asymptotic properties of the up (u) and down (d) parameters, elevating the relative importance of the mean in valuing options. This property of their skewness model suggests that when distributions of logarithmic returns are characterized by skewness, the observed pricing biases associated with the B-S model may be due to not only the omission of skewness, but also the mean.

Today, the derivative market for non-stock options (indices, currencies, debt securities, and commodities) is dominated more by options on futures contracts than options on spot securities. The purpose of this paper is to extend the JPM skewness-adjusted binomial model to the pricing of futures options. In addition, given that one of the features of the JPM skewness model for spot options is that skewness changes the asymptotic properties of the u and d parameters, this paper examines in some detail the asymptotic properties of the skewness-adjusted binomial model as it applies to both futures and spot options. Our results show that the skewness model for futures options has similar asymptotic properties as the model for spot options. However, in the case of futures options, the presence of skewness elevates the importance of the mean, as well as the risk-free rate, the asset yield, and other parameters that are defined by the carrying-cost model.

2. BINOMIAL FUTURES OPTIONS PRICING MODEL

The standard binomial option model values futures options recursively by determining the futures option's intrinsic values at expiration and then using the singleperiod binomial model at each node to price the futures option equal to the value of its replicating portfolio:

$$C_0^* = \frac{1}{e^{R(t/n)}}\left[\sum_{j=0}^{n}\frac{n!}{(n-j)!j!}p^j(1-p)^{(n-j)}\right.$$
$$\left.\left[\text{Max}\left(u^{fj}d^{f(n-j)}f_0\right)-X,0\right]\right] \tag{1}$$

$$P_0^* = \frac{1}{e^{R(t/n)}}\left[\sum_{j=0}^{n}\frac{n!}{(n-j)!j!}p^j(1-p)^{(n-j)}\right.$$
$$\left.\left[\text{Max}(X-\left(u^{fj}d^{f(n-j)}f_0\right),0\right]\right] \tag{2}$$

where:

C_0 = call price P_0 = put price R = annual risk-free rate t = time to expiration expressed as a proportion of a year X = exercise price f_0 = current futures price u^f = the futures up parameter d^f = the futures down parameter n = number of periods to the option's expiration t/n = length on the binomial period = time to expiration as a proportion of a year (t) divided by number of periods to the option's expiration p = risk-neutral probability For the case of an option on a financial futures contract (e.g., index, currency, or debt security) in which the underlying security is adjusted to reflect a continuous asset yield (e.g., dividend yield, foreign risk-free rate, or coupon rate), the equilibrium futures price as determined by the carrying-cost model is:

$$f_0 = S_0 e^{(R-\psi)n_f\Delta t} \tag{3}$$

where:

S_0 = current spot price

ψ = annual asset yield

Δt = t/n = length of binomial steps as a proportion of a year n_f = number of discrete binomial periods of length Δt to the futures' expiration The risk-neutral probability, p, for options on financial futures defined in terms of the up and down parameters for the underlying spot ($u = S_u/S_0$ and $d = S_d/S_0$, where $S_u = uS_0$ and $S_d = dS_0$) is:

$$p = \frac{e^{(R-\psi)(t/n)} - d}{u-d} \tag{4}$$

If the carrying-cost model holds, then the up and down parameters for the futures price ($u^f = f_u/f_0$ and $d^f = f_d/f_0$, where $f_u = u^f$ and $f_d = df_0$) are given as:

$$u^f = \frac{f_u}{f_0} = \frac{uS_0e^{(R-\psi)(n_f-1)(t/n)}}{S_0e^{(R-\psi)n_f(t/n)}} = \frac{u}{e^{(R-\psi)(t/n)}}$$

$$u = u^f e^{(R-\psi)(t/n)} \tag{5}$$

$$d^f = \frac{f_d}{f_0} = \frac{dS_0e^{(R-\psi)(n_f-1)(t/n)}}{S_0e^{(R-\psi)n_f(t/n)}} = \frac{d}{e^{(R-\psi)(t/n)}}$$

$$d = d^f e^{(R-\psi)(t/n)} \tag{6}$$

Substituting Equations (5) and (6) into Equation (4), the risk-neutral probability for futures call and put option prices can be alternatively defined in terms of the futures up and down parameters (u^f and d^f) :

$$p = \frac{e^{(R-\psi)(t/n)} - d}{u - d}$$

$$p = \frac{e^{(R-\psi)(t/n)} - d^f e^{(R-\psi)(t/n)}}{u^f e^{(R-\psi)(t/n)} - d^f e^{(R-\psi)(t/n)}}$$

$$p = \frac{1 - d^f}{u^f - d^f} = p^f \qquad (7)$$

2.1. Skewness-Adjusted Formulas for u^f and d^f

In their seminal 1979 paper, Cox, Ross, and Rubinstein [11] derive the formulas for estimating the u and d parameters of the BOPM for a spot option. They do this by setting the equations for the expected value and variance of the logarithmic return of the underlying security equal to their empirical values. The resulting equations are then solved simultaneously for u and d under the assumption that the probability of the underlying security increasing in one period (q) is 0.5. By treating q as an unknown, JPM extend the CRR binomial model to include skewness. Specifically, in the JPM skewness-adjusted model, the u, d, and q values that define a binomial process for a spot security are found by setting the equations for the expected value, variance, and skewness equal to their respective empirical values and then solving the resulting equation system simultaneously for u, d, and q. That is:

$$E(g_n) = \sum_{j=0}^{n} p_{nj}\, g_{nj} = n\left[q \ln u + (1-q)\ln d\right] = t\,\mu_S$$

$$V(g_n) = nq(1-q)\left[\ln(u/d)\right]^2 = t V_S$$

$$Sk(g_n) = n\left[q(1-q)^3 - q^3(1-q)\right]\left[\ln(u/d)\right]^3 = t\,\delta_S, \qquad (8)$$

where:

g = logarithmic return of the underlying spot price = $\ln(S_1/S_0)$

j = number of increases in n periods p_{nj} = probability of j increases in n periods, where:

m_S, V_S, σ_S, d_S = the annualized empirical values of the mean, variance, standard deviation, and skewness of the spot price's logarithmic return t = time to expiration as a proportion of a year n = number of periods The values of u, d, and q that satisfy this system of equations are:

$$u = e^{\sigma_S \sqrt{((1-q)/q)(t/n)} + \mu_S (t/n)} \tag{9}$$

$$d = e^{-\sigma_S \sqrt{(q/(1-q))(t/n)} + \mu_S (t/n)} \tag{10}$$

$$q = \frac{1}{2} \pm \frac{1}{2}\left[\frac{4(tV_S)^3}{n\delta_s^2} + 1\right]^{-1/2}, \quad -\text{ if } \delta_S > 0, + \text{ if } \delta_S < 0. \tag{11}$$

For futures options, the formulas for estimating u^f, d^f, and q are found by setting the equations for the population moments for the futures price's logarithmic return equal to their respective empirical values, and then solving the resulting equation system simultaneously for u^f, d^f, and q:

$$\begin{aligned} &n\left[q \ln u^f + (1-q)\ln d^f\right] = t\mu_f \\ &nq(1-q)\left[\ln\left(u^f/d^f\right)\right]^2 = tV_f \\ &n\left[q(1-q)^3 - q^3(1-q)\right]\left[\ln\left(u^f/d^f\right)\right]^3 = t\delta_f, \end{aligned} \tag{12}$$

where m_f, V_f, σ_f, and d_f are respectively the annualized empirical values of the mean, the variance, the standard deviation, and skewness of the futures price's logarithmic return, $\ln(f_1/f_0)$, and t is the time to expiration as a proportion of a year.

The values of u^f, d^f, and q that satisfy this system of equations are:

$$u^f = e^{\sigma_f \sqrt{((1-q)/q)(t/n)} + \mu_f (t/n)} \tag{13}$$

$$u^f = e^{\sigma_f \sqrt{((1-q)/q)(t/n))} + \mu_f (t/n)} \tag{14}$$

$$q = \frac{1}{2} \pm \frac{1}{2}\left[\frac{4(tV_f)^3}{n(t\delta_f)^2} + 1\right]^{-1/2}, \quad -\text{ if } \delta_f > 0, + \text{ if } \delta_f < 0 \tag{15}$$

2.2. Relations between Futures and Spot Mean and Volatility

The relationships between the spot and futures parameters follow directly from the futures and spot relation defined by the carrying-cost model given in equation (3). Specifically, from equation (3), one can solve the relationship between the spot and futures moments as follows:

$$S_0 = f_0 e^{-(R-\psi)n_f \Delta t}$$

$$S_1 = f_1 e^{-(R-\psi)(n_f - 1)\Delta t}$$

$$\frac{S_1}{S_0} = \frac{f_1 e^{-(R-\psi)(n_f - 1)\Delta t}}{f_0 e^{-(R-\psi)n_f \Delta t}}$$

$$\ln\left(\frac{S_1}{S_0}\right) = \ln\left(\frac{f_1 e^{-(R-\psi)(n_f - 1)\Delta t}}{f_0 e^{-(R-\psi)n_f \Delta t}}\right)$$

$$\ln\left(\frac{S_1}{S_0}\right) = \ln\left(\frac{f_1}{f_0}\right) + \ln\left(e^{(R-\psi)\Delta t}\right)$$

$$\ln\left(\frac{S_1}{S_0}\right) = \ln\left(\frac{f_1}{f_0}\right) + (R-\psi)\Delta t \tag{16}$$

Taking the expected value of both sides of Equation (16) results in the mean over a period of time Δt:

$$E\left[\ln\left(\frac{S_1}{S_0}\right)\right] = E\left[\ln\left(\frac{f_1}{f_0}\right) + (R-\psi)\Delta t\right]$$
$$\mu_S^* = \mu_F^* + (R-\psi)\Delta t$$

where μ_S^* and μ_F^* are respectively the periodic means of a spot option and a futures option for a period of length Δt(= t/n). Multiplying both sides by 1/Δt results in the annualized means ($\mu_S$and μ_F):

$$\mu_S^*(1/\Delta t) = \mu_f^*(1/\Delta t) + (R-\psi)\Delta t$$
$$\mu_S = \mu_f + (R-\psi)$$

Taking the variance of both sides of (16) and annualizing we obtain:

$$Var\left[\ln\left(\frac{S_1}{S_0}\right)\right] = Var\left[\ln\left(\frac{f_1}{f_0}\right) + (R-\psi)\Delta t\right]$$
$$\sigma_S^2 = \sigma_F^2$$
$$\sigma_S = \sigma_F$$

A similar proof shows that $\delta_S = \delta_F$.

Thus, the relationships between the mean, standard deviation, and skewness on the futures and spot logarithmic returns are:

$$\mu_f = \mu_s - (R-\psi) \tag{17}$$

$$\sigma_f = \sigma_S \tag{18}$$

$$\delta_f = \delta_S \tag{19}$$

Substituting (17) and (18) into equations (13) and (14), u^f and d^f can be expressed in terms of the spot mean and variability, the risk-free rate, and the asset yield:

$$\begin{aligned} u^f &= e^{\sigma_f \sqrt{((1-q)/q)(t/n)} + \mu_f (t/n)} \\ &= e^{\sigma_S \sqrt{((1-q)/q)(t/n)} + (\mu_S - (R-\psi))(t/n)} \end{aligned} \tag{20}$$

$$\begin{aligned} d^f &= e^{-\sigma_f \sqrt{((1-q)/q)(t/n)} + \mu_f (t/n)} \\ &= e^{-\sigma_S \sqrt{(q/(1-q))(t/n)} + (\mu_S - (R-\psi))(t/n)} \end{aligned} \tag{21}$$

The difference between the futures up and down parameters (u^f and d^f respectively), and the spot up and down parameter (u and d respectively (see Equations (9) and 10)) is the net cost of carry term (R – ψ). If R > ψ, then the futures market is normal with the futures price (equation (3)) exceeding the spot.

In this market, u > u^f and d > d^f. On the other hand, if R < ψ, then the futures market is inverted with the futures price less than the spot. In this market, u^f > u and d^f > d. Finally, if the net carry cost is zero, then R = ψ and the futures market is neutral; in this market, the futures up and down parameters will be equal to the spot up and down parameters, that is, $u^f = u$ and $d^f = d$. The relations between the parameters are illustrated in **Table 1**. The table shows the futures and spot up and down parameters calculated for a number of scenarios: positive mean and negative skewness cases (increasing price case) for normal, inverted, and neutral futures market; negative mean and positive skewness cases (decreasing price case) for normal, inverted, and neutral futures market; and zero mean and skewness case (stable price case) for normal, inverted, and neutral futures market. Panel A in the table shows the inputs for each scenario and Panel B gives the corresponding parameters.

3. BINOMIAL FUTURES OPTIONS PRICING MODEL

3.1. Decreasing Exchange-Rate Case

Several empirical studies have shown that periods of increasing security prices are often characterized by a positive mean and significant negative skewness in the security price's logarithmic return. As an example of a decreasing price scenario, suppose the current US dollar/British pound exchange rate is S_0 = \$1.60/BP, and there is a market expectation of a dollar appreciation over the next year such that the expected distribution of logarithmic returns for the exchange rate has the following annualized mean, variance, and skewness: $\mu_S = -0.17597$, $V_S = 0.019555$, and $\delta_s = 0.0008602$. Given these empirical moment values, consider the pricing of call and put options on a British pound (BP) futures contract each with X = \$1.60/BP and a time to expiration of 270 days, using a three-period binomial model. In pricing the options, assume the following:

1) The spot \$/BP exchange rate at time 0 is E_0 = \$1.60/BP 2) The annual risk-free rate paid on US dollars is R = 0.06 3) The annual risk-free rate paid on British pounds is $\psi = 0.04$ 4) The futures contract on the BP expires in one year 5) Carrying-cost model holds 6) Options on the BP futures options expire in 270 days 7) 360-day year The length of the binomial period in years is $\Delta t = t/n = (270/360)/3 = 0.25$, with the call and put options expiring in $(n_{option})\Delta t = (3)\ (0.25) = 0.75$ years, and the BP futures expiring in $n_f \Delta t = (4)\ (0.75) = 1$ year. The up and down parameters for the spot rate are u = 1.052636 and d = 0.9091, the up and down parameters for the futures contract are $u^f = 1.047388$ and $d^f = 0.90457$, q = 0.35, the risk-neutral probability is $p^f = 0.66822$, and the equilibrium futures price is \$1.63232/BP:

$$q = \frac{1}{2} \pm \frac{1}{2}\left[\frac{4\left(t\sigma_e^2\right)^3}{n\left(t\delta_e\right)^2} + 1\right]^{-1/2}$$

$$= \frac{1}{2} - \frac{1}{2}\left[\frac{4\left((0.75)(0.019556)\right)^3}{(3)\left((0.75)(0.00086002)\right)^2} + 1\right]^{-1/2}$$

$$= 0.35$$

$$u^f = e^{\sqrt{V_S((1-q)/q)(t/n)} + ((\mu_S - (R-\psi))(t/n)}$$

$$= e^{\sqrt{0.019555((1-0.35)/0.35)(0.75/3)} + (-0.17597-(0.06-0.04))(0.75/3)}$$

$$= 1.047388$$

$$d^f = e^{-\sqrt{V_S(q/(1-q))(t/n)} + (\mu_S - (R-\psi))(t/n)}$$

$$= e^{-\sqrt{0.019555(0.35/(1-0.35))(0.75/3)} + (-0.17597-(0.06-0.04))(0.75/3)}$$

$$= 0.90457$$

$$p^f = \frac{1-d^f}{u^f - d^f} = \frac{1-0.90457}{1.047388-0.90457}$$

$$= 0.66822$$

$$f_0 = S_0 e^{(R-\psi)n_f \Delta t}$$

$$f_0 = \$1.60e^{(.06-.04)4(.25)} = \$1.63232$$

The negative mean and positive skewness in this case yield u^f, d^f, and q values that reflect a decreasing exchange-rate scenario in which the proportional decrease in the futures rate each period is 10.03%, exceeding in absolute value the proportional increase of 4.63%, and the probability of the decrease in each quarterly period ($\Delta t = 0.25$) is $1 - q = 0.65$:

Table 1. u and d and u^f and d^f values for different scenarios.

Panel: A							
Scenario	Market	R	y	t/n	Annualized Mean	Annualized Variance	Annualized Skewness
Positive mean, Zero Skewness	Normal	0.06	0.04	0.25	0.0880	0.0215	0.0000
Positive mean, Negative Skewness	Normal	0.06	0.04	0.25	0.1760	0.0196	−0.0009
Positive mean, Negative Skewness	Normal	0.06	0.04	0.25	0.2640	0.0138	−0.0012
Positive mean, Zero Skewness	Inverted	0.04	0.06	0.25	0.0880	0.0215	0.0000
Positive mean, Negative Skewness	Inverted	0.04	0.06	0.25	0.1760	0.0196	−0.0009
Positive mean, Negative Skewness	Inverted	0.04	0.06	0.25	0.2640	0.0138	−0.0012
Positive mean, Zero Skewness	Stable	0.06	0.06	0.25	0.0880	0.0215	0.0000
Positive mean, Negative Skewness	Stable	0.06	0.06	0.25	0.1760	0.0196	−0.0009
Positive mean, Negative Skewness	Stable	0.06	0.06	0.25	0.2640	0.0138	−0.0012
Negative mean, Zero Skewness	Normal	0.06	0.04	0.25	−0.0880	0.0215	0.0000
Negative mean, Positive Skewness	Normal	0.06	0.04	0.25	−0.1760	0.0196	0.0009
Negative mean, Positive Skewness	Normal	0.06	0.04	0.25	−0.2639	0.0138	0.0012
Negative mean, Zero Skewness	Inverted	0.04	0.06	0.25	−0.0880	0.0215	0.0000
Negative mean, Positive Skewness	Inverted	0.04	0.04	0.25	−0.1760	0.0196	0.0009
Negative mean, Positive Skewness	Inverted	0.04	0.06	0.25	−0.2639	0.0138	0.0012
Negative mean, Zero Skewness	Stable	0.06	0.06	0.25	−0.0880	0.0215	0.0000
Negative mean, Positive Skewness	Stable	0.06	0.06	0.25	−0.1760	0.0196	0.0009
Negative mean, Positive Skewness	Stable	0.06	0.06	0.25	−0.2639	0.0138	0.0012
Zero Mean, Zero Skewness	Normal	0.06	0.04	0.25	0.0000	0.0363	0.0000
Zero Mean, Zero Skewness	Inverted	0.04	0.06	0.25	0.0000	0.0363	0.0000
Zero Mean, Zero Skewness	Stable	0.06	0.06	0.25	0.0000	0.0363	0.0000

Panel B:

Scenario	Market	q	Spot u	Spot d	Futures u^f	Futures d^f
Positive mean, Zero Skewness.	Normal	0.5000	1.1000	0.9500	1.0945	0.9453
Positive mean, Negative Skewness.	Normal	0.6500	1.1000	0.9500	1.0945	0.9453
Positive mean, Negative Skewness.	Normal	0.8000	1.1000	0.9500	1.0945	0.9453
Positive mean, Zero Skewness.	Inverted	0.5000	1.1000	0.9500	1.1055	0.9458
Positive mean, Negative Skewness.	Inverted	0.6500	1.1000	0.9500	1.1055	0.9458
Positive mean, Negative Skewness.	Inverted	0.8000	1.1000	0.9500	1.1055	0.9458
Positive mean, Zero Skewness	Stable	0.5000	1.1000	0.9500	1.1000	0.9500
Positive mean, Negative Skewness	Stable	0.6500	1.1000	0.9500	1.1000	0.9500
Positive mean, Negative Skewness	Stable	0.8000	1.1000	0.9500	1.1000	0.9500
Negative mean, Zero Skewness.	Normal	0.5000	1.0526	0.9091	1.0474	0.9046
Negative mean, Positive Skewness.	Normal	0.3500	1.0526	0.9091	1.0474	0.9046
Negative mean, Positive Skewness.	Normal	0.2000	1.0526	0.9091	1.0474	0.9046
Negative mean, Zero Skewness.	Inverted	0.5000	1.0526	0.9091	1.0579	0.9137
Negative mean, Positive Skewness.	Inverted	0.3500	1.0526	0.9091	1.0526	0.9091
Negative mean, Positive Skewness.	Inverted	0.2000	1.0526	0.9091	1.0579	0.9137
Negative mean, Zero Skewness	Stable	0.5000	1.0526	0.9091	1.0526	0.9091
Negative mean, Positive Skewness	Stable	0.3500	1.0526	0.9091	1.0526	0.9091
Negative mean, Positive Skewness	Stable	0.2000	1.0526	0.9091	1.0526	0.9091
Zero Mean, Zero Skewness.	Normal	0.5000	1.1000	0.9091	1.0945	0.9046
Zero Mean, Zero Skewness.	Inverted	0.5000	1.1000	0.9091	1.1055	0.9137
Zero Mean, Zero Skewness	Stable	0.5000	1.1000	0.9091	1.1000	0.9091

$$\left|\ln d^f\right| > \left|\ln u^f\right|$$
$$\left|\ln 0.90457\right| > \left|\ln 1.047388\right|$$
$$\left|10.03\%\right| > \left|4.63\%\right|$$

The binomial tree for the underlying spot \$/BP exchange rate, BP futures contract, European and American futures calls, and European and American futures puts are shown in Exhibit 1. In the three-period option case, the binomial model prices the European futures call at \$0.0857 and the European futures put at \$0.0553. As shown in the exhibit, there is an early exercise advantage for the American futures call at the upper node in period 2, and an early exercise advantage for the American futures put at the lower node in period 2. As a result, both the American futures put and call options are price slightly higher than their European counterparts.

If the up and down parameters are not adjusted for skewness, then q = 0.5 and the skewness-adjusted equations for the up and down parameters for the spot and futures simplify to the CRR/RB formulas:

$$u = e^{\sigma_S\sqrt{t/n} + \mu_S(t/n)}$$
$$d = e^{-\sigma_S\sqrt{t/n} + \mu_S(t/n)}$$
$$u^f = e^{\sigma_f\sqrt{t/n} + \mu_f(t/n)}$$
$$= e^{\sigma_S\sqrt{(t/n)} + (\mu_S - (R-\psi))(t/n)}$$
$$d^f = e^{-\sigma_f\sqrt{t/n} ++ \mu_f(t/n)}$$
$$= e^{-\sigma_S\sqrt{(t/n)} + (\mu_S - (R-\psi))(t/n)}$$

In this example, the up and down parameters for the spot rate would be u = 1.026268 and d = 0.892335, the up and down parameters on the futures contract would be u^f = 1.021149 and d^f = 0.887884, and the risk-neutral probability would be p^f = 0.8412978. In this case, in which skewness is assumed to exist but is excluded in the estimates of the up and down parameters the binomial model prices the European call at \$0.0786, 8.28% less than the skewness-adjusted model, and the European put at \$0.0477, 13.74% less than the skewness model. Additionally, there would not be an early exercise advantage for the American futures call and there would be only one early exercise advantage for the American futures put. As a result, the American futures put option is priced slightly higher than its European counterpart.

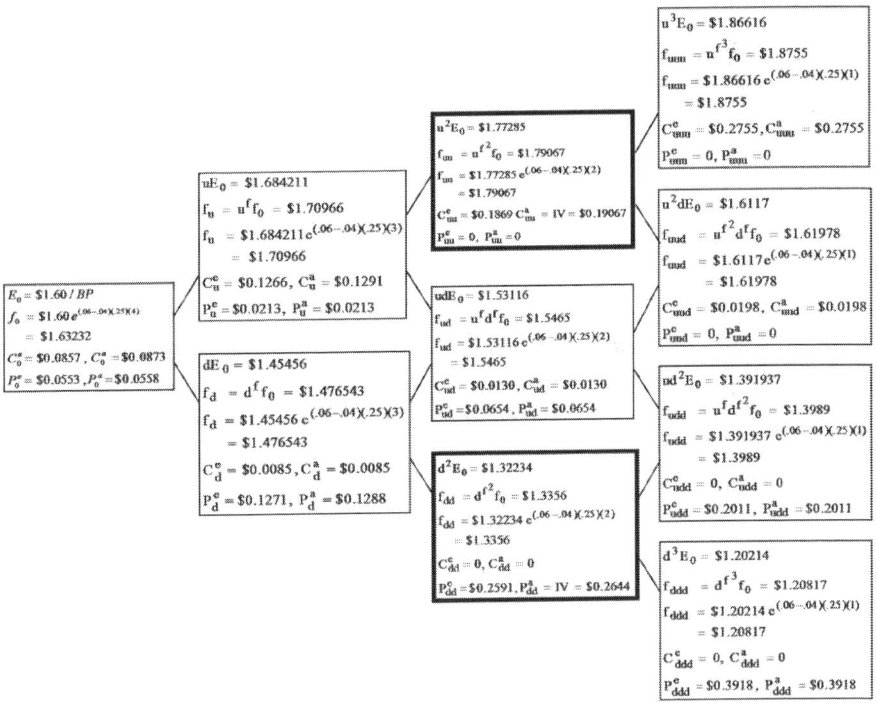

Exhibit 1. Binomial call and put values under decreasing exchange rate case: skewness model.

3.2. Increasing Exchange Rate Case

Periods of increasing security prices are often characterized by a positive mean and negative skewness in the security price's logarithmic return. To illustrate option pricing under this scenario, suppose the market expects the \$/BP exchange rate to increase over the next year such that the expected distribution of the exchange rate's logarithmic return has the following estimated annualized moments: $\mu_S = 0.17597$, $V_S = 0.019555$, and

$\delta_S = -0.0008602$. In this increasing exchange rate case, the up and down parameters for the spot rate are u = 1.10 and d = 0.95, the up and down parameters on the futures contract are $u^f = 1.0945$ and $d^f = 0.94526$, q = 0.65, and the risk-neutral probability is $p^f = 0.36675$. The positive mean and negative skewness yield u^f, d^f, and q values that reflect an increasing exchange-rate period in which the proportional increase in the futures rate each period is 9.03%, exceeding in absolute value the proportional decrease of 5.63%, and the probability of an increase in each quarterly period ($\Delta t = 0.25$) is q = 0.65:

$$|\ln u^f| > |\ln d^f|$$

$$|\ln 1.0945| > |\ln 0.94526|$$

$$|9.03\%| > |5.63\%|$$

In a three-period option case, the binomial model for the increasing case would price the European futures call at $0.0862 and the European futures put at $0.0553. The American futures put and call options are priced slightly higher than the European option, given early exercise advantages for both the call and put options.

If the up and down parameters for this increasing case were not adjusted for skewness, then q would be equal to 0.5, and the up and down parameters for the spot rate would be u = 1.120667 and d = 0.974407, the up and down parameters on the futures contract would be $u^f = 1.115078$ and $d^f = 0.969547$, and the risk-neutral probability would be $p^f = 0.20925$. In this case in which skewness is assumed to exist but is excluded in the estimates of the up and down parameters, the binomial model prices the European call at $0.084, 2.55% less than the skewness-adjusted model, and the European put at $0.0531, 3.98% less than the skewness model.

4. PROPERTIES OF THE SKEWNESS MODEL

In the case of spot options, the existence of skewness affects the relative contribution of the mean to the values of the up and down parameters and the asymptotic properties, with the skewness-adjusted up and down parameters having different asymptotic properties than the CRR/RB parameters. In the case of futures options, these same properties also hold. In addition, with futures options, the impact skewness has on elevating the importance of the mean depends on the carrying cost value (R – ψ) and whether the futures market is normal, inverted or neutral.

4.1. Relative Importance of the Mean Term

In the case of a positive mean, the mean becomes more important in determining the value of the spot up parameter value, the greater the negative skewness (or equivalently the more q exceeds 0.5). Similarly, for futures options, the μ – (R – ψ) term becomes more important in determining the futures up parameter value (u^f), the greater the negative skewness. By contrast, in the case of a negative mean, the mean becomes more important in determining the value of d, and the μ – (R – ψ) term becomes more important in determining the value of the futures down parameter (d^f), the greater the positive skewness (or equivalently the more 1 – q exceeds 0.5).

To see the impact skewness has on increasing the importance of the mean for spot options and the μ – (R – ψ) term for futures options, consider the previous threeperiod increasing exchange rate case in which u = 1.10 and d = 0.95 for the spot exchange rate. If skewness were zero, then q would be equal to 0.5, the expected quarterly mean would be equal to 0.020084 (=[(0.5)ln(1.10) + (0.5)ln(0.95)]) and the annualized mean would be 0.088034. The quarterly variance would be equal to 0.0053731 (= 0.5[ln (1.10) – 0.020084]^2 + 0.5[ln (0.95) – 0.020084]^2), and the annualized variance would be 0.021493. If these were the actual empirical values of μ_S and V_S, then ln(u) would be equal to 0.0953 (u = 1.10), and the mean would contribute 23% to the value of u and the variance would contribute 77% to the value of u:

$$\ln u = \mu_S(t/n) + \left[V_S((1-q)/q)(t/n)\right]^{1/2}$$

$$1 = \frac{\mu_S(t/n)}{\ln u} + \frac{\left[V_S((1-q)/q)(t/n)\right]^{1/2}}{\ln u}$$

$$1 = \frac{0.088034(0.75/3)}{\ln(1.10)}$$

$$+\frac{\left[0.021492((1-0.5)/0.5)(0.75/3)\right]^{1/2}}{\ln(1.10)}$$

$$1 = 0.23 + 0.77$$

For the futures option, the ln(u^f) would be equal to 0.0903095 (u^f = 1.094513) given the net cost of carry of 0.02 (=R – ψ = 0.06 – 0.04). The mean minus the net cost of carry (R – ψ) would, in turn, contribute 19% to the value of u^f and the variance would contribute 81% to the value of u^f:

$$\ln u^f = \left[\mu_S - (R-\psi)\right](t/n) + \left[V_S\left((1-q)/q\right)(t/n)\right]^{1/2}$$

$$1 = \frac{\left[\mu_S - (R-\psi)\right](t/n)}{\ln u^f} + \frac{\left[V_S\left((1-q)/q\right)(t/n)\right]^{1/2}}{\ln u^f}$$

$$1 = \frac{\left[0.088034 - (0.06-0.04)\right](0.75/3)}{\ln(1.094513)} + \frac{\left[0.021492\left((1-0.5)/0.5\right)(0.75/3)\right]^{1/2}}{\ln(1.094513)}$$

$$1 = 0.19 + 0.81$$

If negative skewness were present such that q = 0.65, then the quarterly mean would be equal 0.044 (= [(0.65) ln(1.10) + (0.35) ln(0.95)]) and the annualized mean would be 0.176. The quarterly variance would, in turn, be equal to 0.0389986 (= 0.65 $[\ln(1.10) - 0.044]^2$ + 0.35 $[\ln(0.95) - 0.044]^2$) and the annualized variance would be 0.019558. Given the mean and variance values and a q = 0.65, the implied skewness would be δ = –0.0008602. If these were the actual empirical values of mean, variance, and skewness, then ln(u) would be equal to the nonskewed value of 0.0953 (u = 1.10), but with q = 0.65, the contribution of the mean to the value of u would be 46% and the contribution of the variance to the value of u would be 54%:

$$\ln u = \mu_S(t/n) + \left[V_S((1-q)/q)(t/n)\right]^{1/2}$$

$$1 = \frac{\mu_S(t/n)}{\ln u} + \frac{\left[V_S((1-q)/q)(t/n)\right]^{1/2}}{\ln u}$$

$$1 = \frac{0.175996(0.75/3)}{\ln(1.10)}$$

$$+ \frac{\left[0.019558((1-0.65)/0.65)(0.75/3)\right]^{1/2}}{\ln(1.10)}$$

$$1 = 0.46 + 0.54$$

For the futures options, the $\ln(u^f)$ would likewise be equal to its non-skewed value of 0.0903095 ($u^f = 1.094513$), but the contribution of the mean and the net cost of carry $(R - \psi)$ term to the value of u^f would be 43%, and the contribution of the variance to the value of u^f would be 57%:

$$\ln u^f = \left[\mu_S - (R-\psi)\right](t/n)$$

$$+ [V_S((1-q)/q)(t/n)]^{1/2}$$

$$1 = \frac{\left[\mu_S - (R-\psi)\right](t/n)}{\ln u^f}$$

$$+ \frac{\left[V_S((1-q)/q)(t/n)\right]^{1/2}}{\ln u^f}$$

$$1 = \frac{\left[0.175996 - (0.06-0.04)\right](0.75/3)}{\ln(1.094513)}$$

$$+ \frac{\left[0.019558((1-0.65)/0.65)(0.75/3)\right]^{1/2}}{\ln(1.094513)}$$

$$1 = 0.43 + 0.57$$

Thus, in the case of a positive mean, negative skewness increases the relative importance of the mean on the up parameters for the spot and futures rates1.

By contrast, in the case of a positive mean, the mean term has an opposite directional impact on d and d^f than the variance term, with its negative impact increasing the greater the negative skewness (or equivalently the more q exceeds 0.5). For example, in the no skewness case for the futures option, $\ln(d^f)$ is equal to −0.056292 ($d^f = 0.945262$). In this case, the $\mu - (R - \psi)$ term would contribute −30% to the value of d^f and the variance term would contribute 130% to the value of d^f. In the skewness case, the $\mu - (R - \psi)$ term would contribute −69% to the value of d^f, whereas the variance would contribute 169% to the value of d^f:

Just the opposite relationships hold in the case of a negative mean with positive skewness.Specifically for a negative mean, the mean becomes more important in determining d and d^f the greater the positive skewness, whereas the mean term has an opposite directional impact on u and u^f than the variance term, with its negative impact on the up parameter increasing the greater the positive skewness. **Table 2** summarizes the relative contributions of the mean and variance terms to the spot and futures up and down parameters for the decreasing exchange-rate case and other scenarios with different levels of skewness and different futures markets. Panel A details the relative contributions for the spot and Panel B details the contributions for the futures.

4.2. Asymptotic Properties

In the CRR/RB model, as the number of subperiods (n) increases, the mean term in the exponent for spot options and the $\mu - (R - \psi)$ term for futures options go to zero faster than the square root term. Thus for large n, the CRR/RB model depends only on the volatility. Moreover, with $\sigma_S = \sigma_f$, the up and down parameters for the futures are equal to the spot, regardless of whether the futures market is normal or inverted:

$$u^f = e^{\sigma_f \sqrt{t/n}} = e^{\sigma_S \sqrt{t/n}} = u$$

$$d^f = e^{-\sigma_f \sqrt{t/n}} = e^{-\sigma_S \sqrt{t/n}} = d$$

Table 2. (a) Relative contributions of the mean and variance terms to up and down parameters; (b) Relative contributions of the mean and variance terms to up and down parameters.

Panel: A		Relative Contributions of			
Scenario	Market	Mean to u	Variance to u	Mean to d	Variance to d
Increasing: Positive mean, Zero Skewness.	Normal	0.2309	0.7691	−0.4291	1.4291
Increasing: Positive mean, Negative Skewness.	Normal	0.4616	0.5384	−0.8578	1.8578
Positive mean, Negative Skewness.	Normal	0.6924	0.3076	−1.2865	2.2865
Increasing: Positive mean, Zero Skewness.	Inverted	0.2309	0.7691	−0.4291	1.4291
Increasing: Positive mean, Negative Skewness.	Inverted	0.4616	0.5384	−0.8578	1.8578
Increasing: Positive mean, Negative Skewness.	Inverted	0.6924	0.3076	−1.2865	2.2865
Increasing: Positive mean, Zero Skewness	Stable	0.2309	0.7691	−0.4291	1.4291
Increasing: Positive mean, Negative Skewness	Stable	0.4616	0.5384	−0.8578	1.8578
Increasing: Positive mean, Negative Skewness	Stable	0.6924	0.3076	−1.2865	2.2865
Decreasing: Negative mean, Zero Skewness.	Normal	−0.4290	1.4290	0.2309	0.7691
Decreasing: Negative mean, Positive Skewness.	Normal	−0.8577	1.8577	0.4616	0.5384
Decreasing: Negative mean, Positive Skewness.	Normal	−1.2864	2.2864	0.6924	0.3076
Decreasing: Negative mean, Zero Skewness.	Inverted	−0.4290	1.4290	0.2309	0.7691
Decreasing: Negative mean, Positive Skewness.	Inverted	−0.8577	1.8577	0.4616	0.5384
Decreasing: Negative mean, Positive Skewness.	Inverted	−1.2864	2.2864	0.6924	0.3076
Decreasing: Negative mean, Zero Skewness	Stable	−0.4290	1.4290	0.2309	0.7691
Decreasing: Negative mean, Positive Skewness	Stable	−0.8577	1.8577	0.4616	0.5384
Decreasing: Negative mean, Positive Skewness	Stable	−1.2864	2.2864	0.6924	0.3076
Stable Market: Zero Mean, Zero Skewness.	Normal	0.0001	0.9999	−0.0001	1.0001
Stable Market: Zero Mean, Zero Skewness.	Inverted	0.0001	0.9999	−0.0001	1.0001
Stable Market: Zero Mean, Zero Skewness	Stable	0.0001	0.9999	−0.0001	1.0001

Panel: B		Relative Contributions of			
Scenario	Market	$\mu - (Rf - \psi)$ to u	Variance to u	$\mu - (Rf - \psi)$ to d	Variance to d
Increasing: Positive mean, Zero Skewness.	Normal	0.1883	0.8117	−0.3021	1.3021
Increasing: Positive mean, Negative Skewness.	Normal	0.4318	0.5682	−0.6928	1.6928
Positive mean, Negative Skewness.	Normal	0.6753	0.3247	−1.0834	2.0834
Increasing: Positive mean, Zero Skewness.	Inverted	0.2692	0.7308	−0.5834	1.5834
Increasing: Positive mean, Negative Skewness.	Inverted	0.4855	0.5115	−1.0584	2.0584
Increasing: Positive mean, Negative Skewness.	Inverted	0.7077	0.2923	−1.5335	2.5335
Increasing: Positive mean, Zero Skewness	Stable	0.2309	0.7691	−0.4291	1.4291
Increasing: Positive mean, Negative Skewness	Stable	0.4616	0.5384	−0.8578	1.8578
Increasing: Positive mean, Negative Skewness	Stable	0.6924	0.3076	−1.2865	2.2865
Decreasing: Negative mean, Zero Skewness.	Normal	−0.5833	1.5833	0.2692	0.7308
Decreasing: Negative mean, Positive Skewness.	Normal	−1.0583	2.0583	0.4885	0.5115
Decreasing: Negative mean, Positive Skewness.	Normal	−1.5333	2.5333	0.7077	0.2923
Decreasing: Negative mean, Zero Skewness.	Inverted	−0.3021	1.3021	0.1883	0.8117
Decreasing: Negative mean, Positive Skewness.	Inverted	−0.8577	1.8577	0.4616	0.5384
Decreasing: Negative mean, Positive Skewness.	Inverted	−1.0833	2.0833	0.6753	0.3247
Decreasing: Negative mean, Zero Skewness	Stable	−0.4290	1.4290	0.2309	0.7691
Decreasing: Negative mean, Positive Skewness	Stable	−0.8577	1.8577	0.4616	0.5384
Decreasing: Negative mean, Positive Skewness	Stable	−1.2864	2.2864	0.6924	0.3076
Stable Market: Zero Mean, Zero Skewness.	Normal	−0.0553	1.0553	0.0498	0.9502
Stable Market: Zero Mean, Zero Skewness.	Inverted	0.0499	0.9501	−0.0554	1.0554
Stable Market: Zero Mean, Zero Skewness	Stable	0.0001	0.9999	−0.001	1.0001

Using theses up and down parameters, the BOPM for futures options converges to the seminal Black futures option model [13] as n gets large.

The skewness-adjusted up and down parameters, though, do not have the same asymptotic properties as the CRR/ RB parameters. In the skewness model, equation (20) for u^f includes a(1 – q)/q term, and equation (21) for d^f includes a q/(1 – q) term, both of which change the order of magnitude as n gets large. Specifically, for the case of negative skewness, the (1 – q)/q term can be rewritten as:

$$\frac{1-q}{q}=\frac{4(V_f)^3}{2n(\delta_f)^2+4(V_f)^3+2\sqrt{n(\delta_f)^2+4(V_f)^3}\sqrt{(\delta_f)^2}} \quad (22)$$

The expression (1 – q)/q in equation (22), in turn, is the same order of magnitude as 1/n. This can be seen by observing that the term [(1 – q)/q]/[1/n] approaches the constant $(V_f)^3/(\delta_f)^2$ as n gets large. That is:

$$\frac{(1-q)/q}{1/n}=$$

$$\frac{4(V_f)^3}{2(\delta_f)^2+\left[4(V_f)^3\right]/n+2\sqrt{(\delta_f)^2+\left[4(V_f)^3/n\right]}\sqrt{(\delta_f)^2}} \quad (23)$$

and in the limit:

$$\frac{(1-q)/q}{1/n}\rightarrow\frac{(V_f)^3}{(\delta_f)^2} \text{ as } n\rightarrow\infty \quad (24)$$

With (1 – q)/q having the same order of magnitude as 1/n, the term $[(1 - q)/q*V_f/n]^{1/2}$ in Equation (13) ap- proaches a constant multiplied by $[V_f]^{1/2}/n$, as n gets large. As a result, for the case of large n, the first term in the exponent in equation (13) approximates a constant divided by n, which is in the same form as the second term, $\mu_f/n = (\mu_S - (R - \psi))/n$.

Consequently, both terms in the exponent of equation (13) for u^f contribute equally, even when n is large. Thus, as n gets large, u^f depends not only on the variance and skewness, but also on the mean, risk-free rate, and asset yield. By contrast, equation (14) for d^f is defined in terms of $q/(1 - q)$. In this case, as n gets large, the variance term in the exponent for d^f approaches the constant δ_f/V_f and the μ_f/n term approaches zero. Thus, for the case of negative skewness with a large n, d^fdepends on the variance and skewness, but not on the mean, risk-free rate and dividend yield, whereas u^f depends on all three parameters. Just the opposite asymptotic relationships occur when skewness is positive. In this case, the mean, variance, skewness, risk-free rate, and dividend yield determine d^f (even when n is large), whereas just the variance and skewness determine u^f.

Note, similar asymptotic relations also hold for the spot up and down parameters, with the spot and futures asymptotic relations being equivalent when the net cost of carry is zero $(R - \psi = 0)$.

To illustrate the relative importance of the mean for the case of large n, consider the preceding increasing exchange-rate case (option expiring in 0.75 year, R = 0.06 and ψ = 0.04) where the estimated annualized mean, variance, and skewness were μ_S = 0.17597, V_S = 0.0019555, and δ_S= −0.0008602. If we subdivide the option period (t = 0.75) into 3 subperiods (n = 3), then for the spot exchange rate the mean term would contribute 46.164% to the value of u and the variance term would contribute 53.836% to the value of u; for the futures, the $\mu - (R - \psi)$ term would contribute 43.18% to the value of u^f and the variance term would contribute 56.82% to the value of u^f:

Spot:

$$\mu_S(t/n)/\ln(u) = 0.17597(0.75/3)/\ln(1.10)$$
$$= 0.46164$$

Futures:

$$[\mu_S - (R-\psi)](t/n)/\ln(u^f)$$
$$= [0.17597 - (0.06 - 0.04)] * (0.75/3)/\ln(1.094513)$$
$$= 0.4318$$

If we subdivide the option period into nine monthly subperiods (n = 9; t/n = 0.75/9 = 0.08333; u = 1.03940, d = 0.9481; u^f = 1.037673, d^f = 0.9465), the mean term would contribute 37.95% and the variance term would contribute 62.05% for the spot rate, and for the futures, the mean term would contribute 35.15% and the variance term would contribute 64.85%:

Spot:

$$\mu_S(t/n)/\ln(u)=0.17599586(0.75/9)/\ln(1.0394)$$
$$=0.3795$$

Futures:

$$[\mu_S-(R-\psi)](t/n)/\ln(u^f)$$
$$=[0.17599586-(0.06-0.04)]*(0.75/9)/\ln(1.037673)$$
$$=0.3515$$

For larger n, the contribution of the mean is approximately the same: when n = 39 (weekly periods; t/n = 0.75/39 = 0.01923), the relative contribution of the mean would be 31.59% for the spot rate and 29.04% for the futures; when n = 270 (daily; t/n = 0.75/270 = 0.0028), the contributions would be 28.91% for the spot and 26.49% for the futures; when n = 1000, the contributions would be 28.51% and 26.11% for the spot and futures, respecttively; when n = 1000000, the contributions would be 28.35% and 25.97%.

By contrast, if skewness were zero (the CRR/RB model with $\mu_S = 0.17597$, $V_S = 0.019558$, and $\delta_S = 0$), the relative contributions of the mean term for the spot rate and futures would be 38.62% and 35.80% for n = 3 ((n = 3; t/n = 0.75/9 = 0.08333; u = 1.120667, d = 0.974407; $u^f = 1.115078$, $d^f = 0.969547$):

Spot:

$$\mu_S(t/n)/\ln(u)=0.17599586\ (0.75/3)/\ln(1.120667)$$
$$=0.3862$$

Futures:

$$[\mu_S-(R-\psi)](t/n)/\ln(u^f)$$
$$=[0.17599586-(0.06-0.04)]*(0.75/3)/\ln(1.115078)$$
$$=0.3580$$

When n = 9, the mean contribution for the spot and futures, respectively, would be 26.64% and 24.36%; when n = 39, 14.86% and 14.40%; when n = 270, 6.22% and 5.55%; when n = 1000, 3.33% and 2.96%; and when n = 1000000, 0.1% and 0.1%.

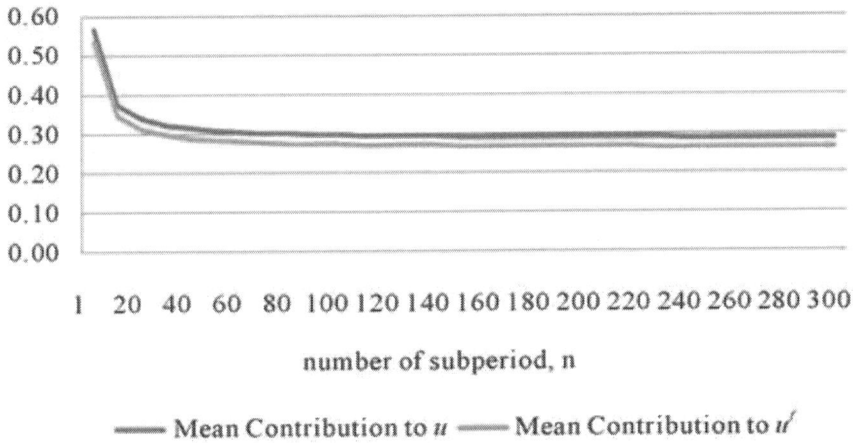

Figure 1. Mean contribution to u and u^f, $\mu_S = 0.175996$, $V_S = 0.019558$, $\delta_S = -0.0086019$.

Figure 1 shows graphically the relationship between the number of subperiods and the mean's contribution to the up parameter for the skewness-adjusted increasing exchange-rate case.

The graph in **Figure 1** highlights the asymptotic relation, showing that as n increases, the mean's contribution to both the sopt and futures up parameters decreases asymptotically with the asymptote occurring at approximately $n^* = 30$ where the minimum mean contribution is 30%2.**Figure 2** summarizes the relationship between the number of subperiods and the mean's contribution to the up and down parameters for the CRR/RB case in which skewness is zero.

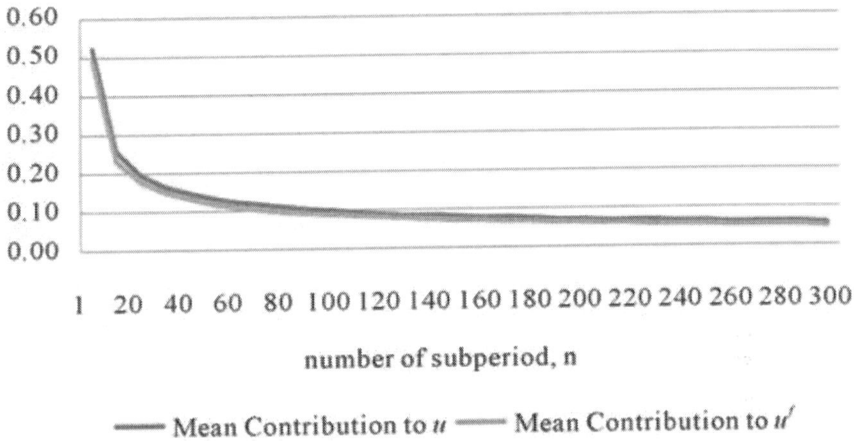

Figure 2. Mean contribution to u and u^f, $\mu_S = 0.175996$, $V_S = 0.019558$, $\delta_S = 0$.

Figure 2, in turn, shows a similar asymptotic relation between the mean's contribution and n for the CRR/RB model as the skewness case, but with the minimum mean contribution being close to zero.

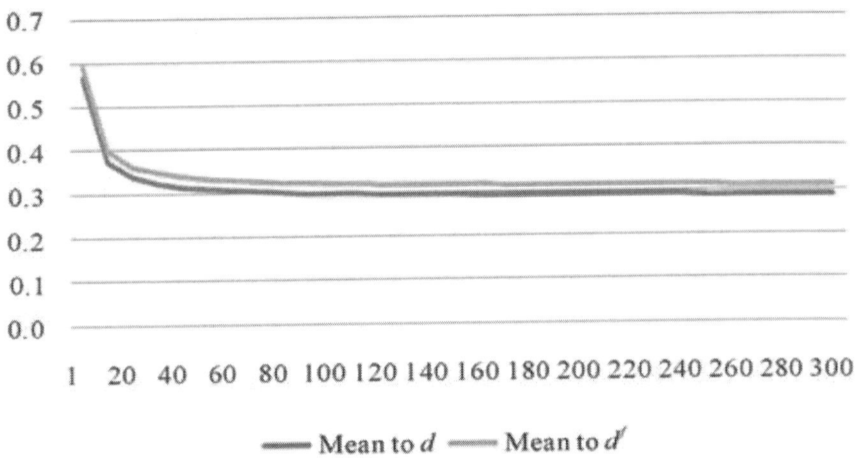

Figure 3. Mean contribution to d and d^f, $\mu_S = -0.17597$, $V_S = 0.019555$, $\delta_S = 0.00086019$.

Figure 3 shows graphically the relationship between the number of subperiods and the mean's contribution to the down parameters for the skewness-adjusted decreasing exchange-rate case characterized by a negative mean and positive skewness: $\mu_S = -0.17597$; $V_S = 0.019555$; $\delta_S = 0.0008602$. The graph highlights the asymptotic relation, showing that as n increases, the mean's contribution to the down parameters decreases asymptotically with the asymptote occurring at approximately $n^* = 30$ where the minimum mean contribution is 30%.

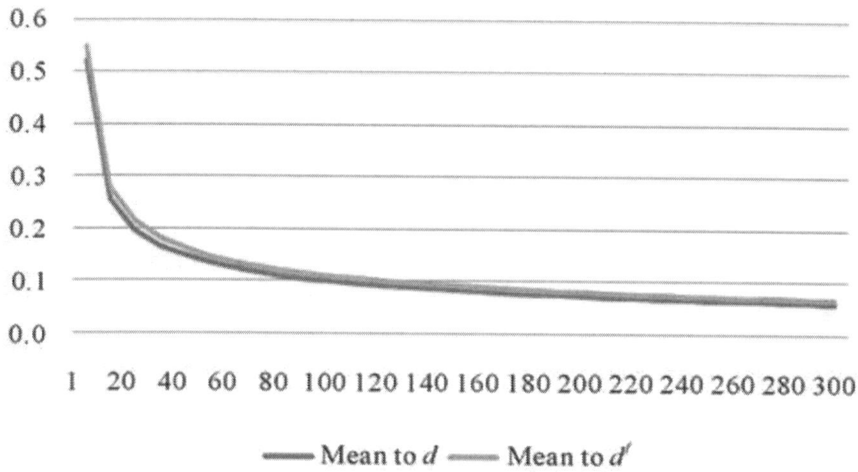

Figure 4. Mean contribution to d and d^f, $\mu_S = -0.17597$, $V_S = 0.019555$, $\delta_S = 0$.

Figure 4 shows a similar asymptotic relation between the mean's contribution to the down parameter and n for the CRR/RB case, but with the minimum contribution being close to zero.

5. DIFFERENCES IN FUTURES OPTION PRICES BETWEEN THE CRR/RB MODEL AND THE SKEWNESS-ADJUSTED MODEL

Values for European futures call options on the British pound obtained using the skewness-adjusted model and the CRR/RB Model are presented in Table 3, and values for the European futures put options on the British pound are shown in Table 4. The call and put futures options each have exercise prices of $1.60/BP and expire in 0.75 years, and the British pound futures contract expires in one year, with the futures price assumed to be equal to its carrying-coat value. The binomial model used for pricing is subdivided into 60 periods of length 6 days. The tables show three futures markets: a normal futures market where the annualized risk-free rate is assumed to be 6% on US dollars and 4% on British pounds, an inverted market where the dollar rate is 4% and British pound rate is 6%, and a neutral market where each rate is equal to 6%. Finally, the tables show two exchange-rate scenarios:

1) An increasing exchange-rate case characterized by a positive mean and a negative skewness: $\mu_S = 0.17599586$, $V_S = 0.019558$, $\delta_S = -0.000860192$.

2) A decreasing exchange-rate case characterized by a negative mean and a positive skewness: $\mu_S = 0.17599586$, $V_S = 0.019558$, $\delta_S = -0$.

A comparison of the futures option values obtained using the skewness-adjusted model with the CRR/RB model illustrates the pricing differences that occur under increasing or decreasing exchange-rate cases characterized by skewness. In general, for both scenarios, the CRR/RB model prices the American and European futures call and the American and European futures puts less than the skewness model, with the greatest pricing differences occurring for out-of-the-money options. Specifically, the si-

mulations show underpricing of the CRR/RB model, ranging from approximately 5% for in-the-money options to approximately 35% for out-of-the-money options. For the futures call option, in both the increasing and decreasing cases, the CRR/RB model underprices more for the inverted futures market case (range 30.56% - 10.11% (increasing); 39.43% - 5.93% (decreasing)) than the normal futures market case (21.59% - 5.63% (increasing); 34.79% - 4.10% (decreasing)). For puts, the CRR/RB model underprices more for the normal market case (range 5.05% - 38.36% (increasing); 9.14% - 30.80% (decreasing)) than the inverted market case (4.67% - 33.06% (increasing); 4.85% - 21.25% (decreasing))3.

Finally, Table 5 compares futures call and put prices with corresponding spot call and put prices for normal, inverted, and neutral futures market under a stable exchange-rate scenario in which the mean and skewness are zero. As shown, with zero skewness, the skewness model and the CRR/RB model for spot and futures options are the same. The simulations, in turn, also show that this is the only case in which European futures and spot options are equal.

Table 3. Comparison of skewness-adjusted futures option model with crr/rb model for normal, inverted and neutral futures markets and increasing and decreasing exchange-rate trends: european futures call options.

Panel A: Normal					
Rate Trend	Spot $/BP	Futures Price	Skewness Model European	CRR/RB Model European	% Difference CRR/RB-Skew
Increase	$1.45	$1.4793	$0.0352	$0.0276	−21.59%
	1.55	1.5813	0.0765	0.0646	−15.56%
	1.65	1.6833	0.1349	0.1217	−9.79%
	1.75	1.7854	0.2078	0.1961	−5.63%
Decrease	$1.45	$1.4793	$0.0411	$0.0268	−34.79%
	1.55	1.5813	0.0797	0.0637	−20.08%
	1.65	1.6833	0.1346	0.1213	−9.88%
	1.75	1.7854	0.2048	0.1964	−4.10%

Panel B: Inverted

Rate Trend	Spot $/BP	Futures Price	Skewness Model European	CRR/RB Model European	% Difference CRR/RB-Skew
Increase	$1.45	$1.4213	$0.0216	$0.0150	−30.56%
	1.55	1.5193	0.0524	0.0397	−24.24%
	1.65	1.6173	0.0993	0.0829	−16.52%
	1.75	1.7153	0.1612	0.1449	−10.11%
Decrease	$1.45	$1.4213	$0.0246	$0.0149	−39.43%
	1.55	1.5193	0.0527	0.0400	−24.10%
	1.65	1.6173	0.0961	0.0838	−12.80%
	1.75	1.7153	0.1552	0.1460	−5.93%

Panel C: Neutral

Rate Trend	Spot $/BP	Futures Price	Skewness Model European	CRR/RB Model European	% Difference CRR/RB-Skew
Increase	$1.45	$1.45	$0.0276	$0.0203	−26.54%
	1.55	1.55	0.0632	0.0508	−19.62%
	1.65	1.65	0.1154	0.1006	−12.82%
	1.75	1.75	0.1823	0.1682	−7.73%
Decrease	$1.45	$1.45	$0.0319	$0.0201	−36.99%
	1.55	1.55	0.0649	0.0506	−22.03%
	1.65	1.65	0.1137	0.1008	−11.35%
	1.75	1.75	0.1779	0.1691	−4.95%

Table 4. Comparison of skewness-adjusted futures option model with crr/rb model for normal, inverted and neutral futures markets and increasing and decreasing exchange-rate trends: european futures put options.

Rate Trend	Spot $/BP	Futures Price	Skewness Model European	CRR/RB Model European	% Difference CRR/RB-Skew
Panel A: Normal					
Increase	$1.45	$1.4793	$0.1506	$0.1430	−5.05%
	1.55	1.5813	0.0944	0.0824	−12.71%
	1.65	1.6833	0.0533	0.0420	−24.05%
	1.75	1.7854	0.0305	0.0188	−38.36%
Decrease	$1.45	$1.4793	$0.1565	$0.1422	−9.14%
	1.55	1.5813	0.0975	0.0816	−16.31%
	1.65	1.6833	0.0549	0.0417	−24.04%
	1.75	1.7854	0.0276	0.0191	−30.80%
Panel B: Inverted					
Rate Trend	Spot $/BP	Futures Price	Skewness Model European	CRR/RB Model European	% Difference CRR/RB-Skew
Increase	$1.45	$1.4213	$0.1950	$0.1859	−4.67%
	1.55	1.5193	0.1307	0.1180	−9.72%
	1.65	1.6173	0.0825	0.0661	−19.88%
	1.75	1.7153	0.0493	0.0330	−33.06%
Decrease	$1.45	$1.4213	$0.1980	$0.1884	−4.85%
	1.55	1.5193	0.1311	0.1184	−9.69%
	1.65	1.6173	0.0793	0.0670	−15.51%
	1.75	1.7153	0.0433	0.0341	−21.25%

Panel C: Neutral

Rate Trend	Spot $/BP	Futures Price	Skewness Model European	CRR/RB Model European	% Difference CRR/RB-Skew
Increase	$1.45	$1.45	0.1710	0.1637	−4.27
	1.55	1.55	0.1110	0.0986	−11.17
	1.65	1.65	0.0676	0.0528	−21.89
	1.75	1.75	0.0389	0.0248	−36.25
Decrease	$1.45	$1.45	0.1753	0.1635	−6.73
	1.55	1.55	0.1127	0.0984	−12.69
	1.65	1.65	0.0659	0.0530	−19.58
	1.75	1.75	0.0345	0.0257	−25.51

Table5. Comparison of european spot and future option prices for normal, inverted and neutral futures markets and a stable exchange-rate market.

	Spot $/BP	Skewness and CRR/RB Spot Option Price European	Futures Price	Skewness and CRR/RB Futures Option Price European	%Difference Futures – Spot Option Price
Panel A: Normal					
Put	$1.45	0.0258	$1.48	$0.0275	6.59%
	1.55	0.0610	1.58	0.0648	6.23%
	1.65	0.1165	1.68	0.1221	4.81%
	1.75	0.1896	1.79	0.1965	3.64%
Call	$1.45	0.1482	$1.48	$0.1429	–3.58%
	1.55	0.0864	1.58	0.0827	–4.28%
	1.65	0.0448	1.68	0.0425	–5.13%
	1.75	0.0209	1.79	0.0192	–8.14%
Panel B: Inverted					
	Spot $/BP	Skewness and CRR/RB Spot Option Price European	Futures Price	Skewness and CRR/RB Futures Option Price European	%Difference Futures – Spot Option Price
Put	$1.45	$0.0166	$1.42	$0.0155	–6.63%
	1.55	0.0429	1.52	0.0405	–5.59%
	1.65	0.0883	1.62	0.0841	–4.76%
	1.75	0.1522	1.72	0.1462	–3.94%
Call	$1.45	$0.1831	$1.42	$0.1889	3.17%
	1.55	0.1138	1.52	0.1188	4.39%
	1.65	0.0636	1.62	0.0673	5.82%
	1.75	0.0320	1.72	0.0343	7.19%

Panel C: Neutral

	Spot $/BP	Skewness and CRR/RB Spot Option Price European	Futures Price	Skewness and CRR/RB Futures Option Price European	%Difference Futures – Spot Option Price
Put	$1.45	$0.0206	$1.45	$0.0206	0.00%
	1.55	0.0510	1.55	0.0510	0.00%
	1.65	0.1010	1.65	0.1010	0.00%
	1.75	0.1692	1.75	0.1692	0.00%
Call	$1.45	$0.1640	$1.45	$0.1640	0.00%
	1.55	0.0988	1.55	0.0988	0.00%
	1.65	0.0532	1.65	0.0532	0.00%
	1.75	0.0258	1.75	0.0258	0.00%

6. CONCLUSIONS

A subtle feature of the B-S model and the BOPM for large n is that these models depend only on the variance. The mean is not important in determining the value of spot options, and the mean and net carry cost are not important for futures options. This feature is a consequence of the assumption that the logarithmic return of the underlying security is normally distributed. In this paper, we show that in cases where skewness exists, the skewness-adjusted up and down parameters for spot options depend more on the mean than non-skewness-adjusted parameters, and that the skewness-adjusted up and down parameters for futures options depend more on the $\mu - (R - \psi)$ term than non-skewness-adjusted parameters. Furthermore, the skewness-adjusted up and down parameters also have different asymptotic properties such that for large n, the mean for spot prices and the $\mu - (R - \psi)$ term for futures maintain their relative importance.

Thus, the presence of skewness serves to augment the relative importance of the mean for spot options and the $\mu - (R - \psi)$ term for futures options. Using simulations, we show that when there is an expected increasing price trend characterized by a positive logarithmic mean and negative skewness or an expected decreasing price trend characterized by a negative logarithmic mean and positive skewness, the CRR/RB model for large n when compared to the skewness-adjusted futures options model underprices futures options between 4% and 30%, with the larger underpricing occurring for out-of-the money options.

Moreover, with the relative contribution of the mean and carrying cost values for a large number of subperiods being at least 30% for one of the up or down parameters in the skewness model and minimal in the CRR/RB, the underpricing can be explained by the change in asymptotic properties resulting from skewness that elevate the importance of the mean and carrying cost parameters. Finally, it should be noted that the BOPM for futures options converges to the seminal Black futures option model as n gets large given a normal distribution for the logarithmic return. The Black futures model, in turn, differs from the B-S Merton model used for pricing spot options by the exclusion of the risk-free rate and the asset yield in the equations for d_1 and d_2. This difference, though, only holds given the assumption of normality. If skewness exists, then the risk-free rate and asset yield, as well as the mean, become important in pricing futures options for the discrete binomial model, as well as the Black futures option model.

REFERENCES

1. R. S. Johnson, J. E. Pawlukiewicz and J. Mehta, "Binomial Option Pricing with Skewed Asset Returns," Review of Quantitative Finance and Accounting, Vol. 9, No. 1, 1997, pp. 89-101.

2. F. Black and M. Scholes, "The Pricing of Options and Corporate Liabilities," Journal of Political Economy, Vol. 81, No. 3, 1973, pp. 637-659.

3. R. S. Johnson, R. A. Zuber and J. M. Gandar, "Binomial pricing of Fixed-Income Securities for Increasing and Decreasing Interest Rate Cases," Applied Financial Economics, Vol. 16, No. 14, 2006, pp. 1029-1046.

4. R. S. Johnson, R. A. Zuber and J. M. Gandar, "Pricing Stock Options under Expected Increasing and Decreasing Price Cases," Quarterly Journal of Business and Economics, Vol. 46, No. 4, 2007, pp. 63-90.

5. E. M. Stein and J. C. Stein, "Stock Price Distributions with Stochastic Volatility: An Analytical Approach," Review of Financial Studies, Vol. 4, No. 4, 1991, pp.727-752.

6. J. B. Wiggins, "Option Values under Stochastic Volatility: Theory and Empirical Estimates," Journal of Financial Economics, Vol. 19, No. 2, 1987, pp. 351-372.

7. S. L. Heston, "A Closed Form Solution for Options and Stochastic Volatility with Applications to Bond and Currency Options," Review of Financial Studies, Vol. 6, No. 2, 1993, pp. 327-344.

8. R. Jarrow and A. Rudd, "Approximate Option Valuation for Arbitrary Stochastic Processes," Journal of Financial Economics, Vol. 10, No. 3, 1982, pp. 347-369.

9. C. J. Corrado and T. Su, "Skewness and Kurtosis in S&P 500 Index Returns Implied by Option Prices," The Journal of Financial Research, Vol. 19, No. 2, 1996, pp. 175- 192.

10. A. Câmara and S. Chung, "Option Pricing for the Transformed-Binomial Class," The Journal of Futures Markets, 2006, Vol. 26, No. 8, pp. 759-787.

11. J. C. Cox, S. A. Ross and M. Rubinstein, "Option Pricing: A Simplified Approach," Journal of Financial Economics, Vol. 7, No. 3, 1979, pp. 229-263.

12. R. J. Rendleman and B. J. Bartter, "Two-State Option Pricing," Journal of Finance, Vol. 34, No. 5, 1979, pp. 1093- 1110.

13. F. Black, "The Pricing of Commodity Contracts," Journal of Financial Economics, Vol. 3, No. 1-2, 1976, pp. 167-179.

NOTES

[1]Note that in this case, $\mu > (R - \psi)$ and the futures market is normal $(R - \psi > 0)$ with the futures price exceeding the spot price. As a result, negative skewness increases the mean's impact on the up parameter (46%) for the spot option more than the up parameter for the futures option (43%). In contrast, if the market were inverted $(R - \psi < 0)$, then negative skewness would have decreased the impact of the mean on the futures up parameter more than the spot parameter.

[2]It should be noted that since all the parameters contribute to either lnu or lnd as n gets large, the n value in which the skewness-adjusted model approaches a continuous one depends on the relative values of μ_S, V_S, and δ_S. For the case of $\delta_S < 0$, the term $(1 - q)/q$ approaches a constant divided by n in the limit. The critical value, n*, can therefore be found by solving for the n that makes $(1 - q)/q$ (Equation (20)) equal to a large proportion (e.g., 0.99) of the limit (Equation (22)). Defining the proportion as $1 - \varepsilon$, where ε is equal to the proportion of error (e.g., $\varepsilon = 0.01$), the n* that is equal to $1 - \varepsilon$ of the limit is:

$$n^* = \frac{(V_e)^3 (1-\varepsilon)\left(1+\sqrt{1+\varepsilon}\right)}{(\delta_e)^2 \varepsilon}$$

[3]The pricing differences between the CRR/RB model and the skewness model are consistent with the aforementioned empirical studies of Stein and Stein [4], Wiggins [5], and Heston [6] who demonstrate that when skewness exists, the B-S model consistently underprices options. Also, as expected, there were no significant difference in the prices for the European call and put options obtained using the Black futures option model (not shown) and the 60-period CRR/RB binomial prices shown in Table 5.

7

CHAPTER

Private and Public Debt Markets in Disequilibrium Theory

Frederick Betz[1,2]

[1]*SUNY Korea, Incheon, South Korea*

[2]*Portland State University, Portland, OR, USA*

ABSTRACT

In disequilibrium pricing of financial markets, excesses in either public debt or private debt can trigger a financial crisis, with attendant bank panics and recessions. For example, in the Euro crisis beginning in 2010, financial contagion in the sovereign bond market has spread among five nations: Greece, Ireland, Cyprus, Portugal, and Spain. But the reasons for the contagion was initially different for the countries, due to either disequilibrium pricing in public debt markets or disequilibrium pricing in private debt markets. In previous papers, we introduced a time-independent- supply-demand model (three-dimensional model) for disequilibrium pricing in financial markets [1] and a steady-state disequilibrium systems model for the run-up of a financial crisis in a public debt market [2] . In this paper we construct a time-dependent disequilibrium systems model for the run-up of financial crises in a private debt market. We analyze the empirical case of the 2010- 12 Euro crisis in Spain (private debt crisis) and then compare this to the empirical case of the 2010-2015 Euro crisis in Greece (public debt crisis).

Keywords: Financial Systems, Fiscal Crisis, Economic Recessions

1. INTRODUCTION

Irvin Minsky proposed a "price disequilibrium theory" for describing the dynamics of financial bubbles [3] ; and the author constructed a graphic depiction of financial transactions in capital-asset markets to depict Minsky's disequilibrium theory [1] . In beginning and first decade of the twenty-first century, financial markets in both private-credit and public-credit experienced disequilibrium instabilities. In 2000, private-credit market instability occurred in the "dot-com bubble" of the U.S. stock market. In 2007, private-credit market instability occurred in the "mortgage-derivative bubble" of the U.S. Wall Street sector, threatening to shut down the U.S. financial system. In 2010, public-credit market instability occurred in a "sovereign-bond bubble" of the Greek bond market.

Figure 1 shows the relative debt levels of public and private debt, as percentages of GDP of the different nations in Europe and North America. Public debt is government debt, as shown in blue on the chart. Private debt is the sum of corporate and household debt as shown in green and brown on the chart.

Greece had the largest public-debt-as-percentage-of-GDP at 127% (PD/GDP), while Italy was close at 116%. The difference between Italy and Greece was tax policy. Greece had continuously run a government budget deficit, due to a reluctance to collect sufficient taxes. Martin Wolf expressed this: "What made the Greek fiscal position so bad was not that its spending was extraordinarily high by Eurozone standards, but rather that was revenue was so low, given the country's high spending" [4] . The Greek tax policy was continually delinquent, never collecting sufficient revenue to fund its government services (in good economic times or bad). Greek governments used foreign borrowing (sovereign bonds) to fund their current expenses, covering perennial budget deficits. In 2009, Ireland at 66% (PD/GDP) and Spain at 53% (PD/GDP) and Portugal at 76% (PD/GDP) were all cases first of private debt market financial bubbles (and were not initially public debt crises).

Yet in these cases of Ireland, Portugal, and Spain, their private debt crises immediately triggered public debt crises. Martin Wolf wrote: "... once (a private sector) crisis hit, the private-sector financial deficit disappeared swiftly, whereas the fiscal deficit soared. The explanation for the latter is that government revenue fell dramatically, and crisis-driven government spending automatically rose, as the banks were rescued... When private spending collapses, as happened in Ireland and Spain, the fiscal deficit always rises more quickly than the external deficit falls ..." [4] . Thus one needs to understand financial crises in both private and public debt markets to understand the Euro crisis of 2010-15.

In the financial contagion of the European sovereign bond market, contagion began with Greek bonds and next ran through Irish, Portuguese, Spanish bonds (while worrying the Italian and French bond market for a time). But the sources of the contagion differed in the different countries. In Greece, it was a financial contagion in the public debt market. In Ireland and Spain, it was contagion in private debt; and in Portugal, a contagion in both private and public debt. The contagions in the national private debt markets were due to housing booms, fueled by cheap Euro dollar credit.

2. PRIVATE DEBT MARKET—DISEQUILIBRIUM SYSTEMS MODEL IN THE CASE OF SPAIN 2010

In the Euro crisis of 2010, two different financial markets, public and private, had triggered Minsky moments in different national financial systems. In Greece, it was in the public debt (sovereign bond market for Greek bonds) that a Minsky moment occurred; but in Spain, the Minsky moment occurred in private debt (housing market). Since 1984, private debt had risen in the Spanish housing market, until in 1999, housing prices began to climb into a "bubble" (as shown in Figure 2).

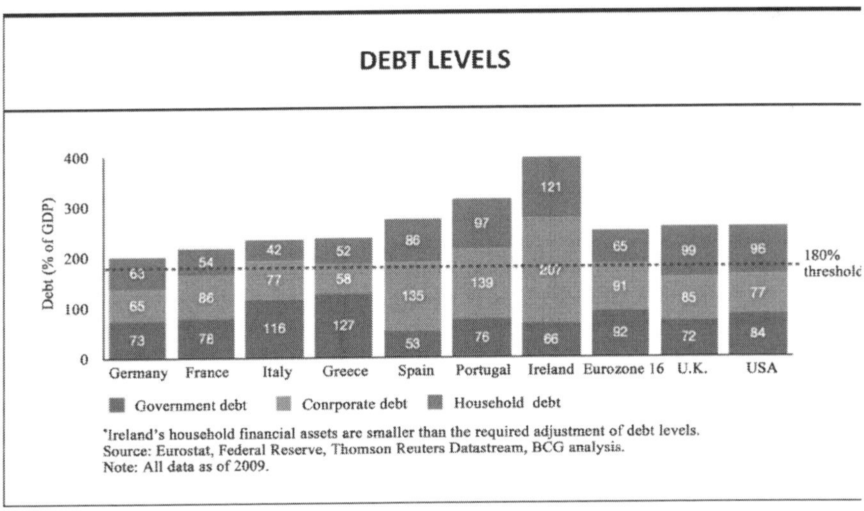

Figure 1. Debt levels in Europe and the USA in 2009. Source:http://en.wikipedia.org/wiki/European_debt_crisis

Residence ownership of houses was encouraged by the government and grew to 80% occupancy. From 2001 through 2005, variable rate interest mortgages were used and property developments increased. Over 500,000 residences were constructed in 2004, with about 300,000 purchased by Spanish citizens and 100,000 by foreign investors. When the housing bubble burst in 2007, housing sales immediately plunged 25%. Housing prices continued to decline and by 2012, had triggered a government crisis.

We next model this financial bubble (disequilibrium-price) in Spain from 2008-10 to 2012. In the Spanish private debt financial market for housing construction, interest rates and leverage were two controls on the flow of loans. Both leverage and interest rates were structural factors which could change over time. Until 2007, interest rates were low as Spanish banks borrowed in the Euro market; but interest rates in Spain jumped, after the 2007 U.S. fiscal bubble burst in U.S. mortgage-derivative market. Since structural features in a financial system change during the speculation in the market leading up to a financial bubble, a disequilibrium systems model must explicitly display a dimension-of-time for the structural change—as shown in Figure 3.

In this temporal model, a Trader obtains a Loan from a Bank with Leverage to purchase a capital asset in a Market in the time period T1. Rates on Interest and also Leverage are both necessary for the Trader to optimize profits, when the trader resells its Capital Asset purchase at the future time. The capital-asset market was in price equilibrium when traders begen speculation in the market, with the reflexive expectation that prices in the market would rose above equilibrium price. Later in the time period T2, bank financing of trader loans changed the market financing from a state of "Speculative" to that of "Ponzi" and then the global financial market collapsed in 2008. The "Minsky moment" occurred when the real estate bubble in Spain "burst" and the Spanish banks financing the real estate loans suddenly became "insolvent".

Next the Spanish government rescued the insolvent Spanish banks, holding defaulted mortgages as collateral. Tom Burridge wrote: "Spain is to get up to 100 bn Euros in loans from Eurozone funds to try to help shore up its struggling banks. The move was agreed during emergency talks with Eurozone finance ministers... Most analysts and experts would say this is a bailout... and the money is going to be directed via the Spanish government using the Fund for Orderly Bank Restructuring (which is a bank restructuring fund in Spain). But it is a government institution, so the debt is going to be on the Spanish government's books. It is money that will be directed towards troubled Spanish banks, banks that lent heavily during the construction boom here—of course, the property market crashed and the loans turned bad" [5] .

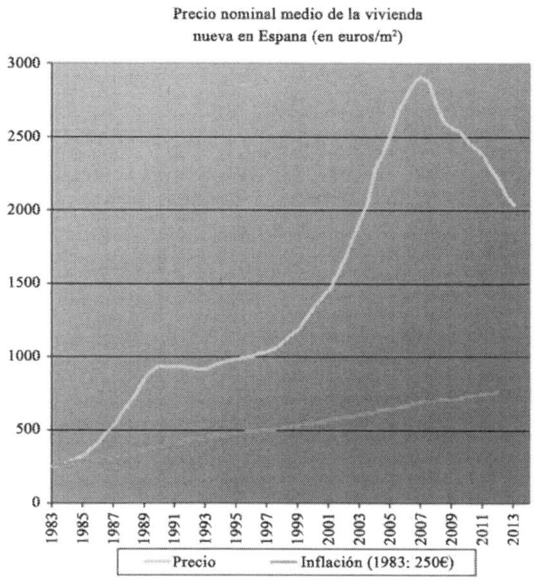

Figure 2. Spanish house prices 1984-2013. Source:http://en.wikipedia.org/wiki/Spanish_property_bubble

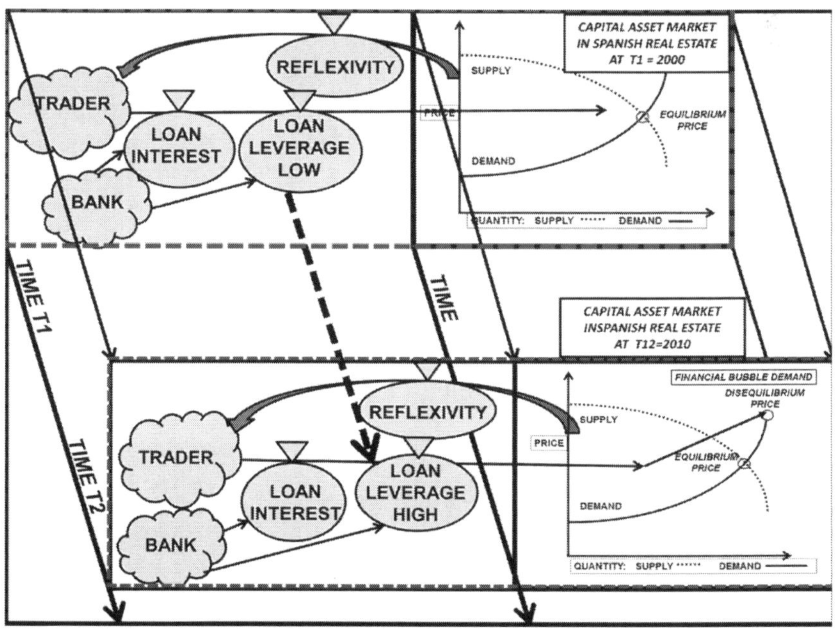

Figure 3. Spanish private debt bubble.

The government rescue of the private banks was necessary because of the insolvency of the banks. Floyd Norris had written: "The Spanish banks are bleeding from loans secured by raw real estate, and from loans for construction. The pain is made worse because such lending soared during the property boom. It is those loans that are now devastating bank balance sheets, as developers who borrowed to build offices, stores and neighborhoods saw demand dry up and now cannot pay the banks back. Just when markets were focused on the risks of a Greek default and the possibility of contagion to other countries, Spain's central bank reported this week that things were getting worse for that country's banks—but not because they held a lot of Greek debt or bonds issued by other troubled economies... With Spain's economy weak and home prices falling, bad loans are growing. And the central bank thinks things are getting worse... It is impossible for Spanish banks to walk away from a collapsing bubble in Spanish real estate... The Bank of Spain has created a program to force mergers of the smaller banks and to bring in better management. It has put about 11 billion Euros into the banks to recapitalize them, and is putting in another 15 billion Euros in a process... (with) an estimate is that the overall number needed is closer to 50 billion Euros" [6] .

The Spanish government provided more than 26 billion Euros to the Bank of Spain to distribute to Spanish banks. The funds were loaned to the government by the "Troika" of the European Union and the International Monetary Fund. In return, the Spanish government agreed to reforms imposed as an austerity policy by the Troika institutions. The Spanish government debt did not soar until after the private debt housing bubble burst, and the government stepped in as "lender-of-last-resort" to save the Spanish banks. We next depict the role of the central bank of Spain in this connection between the financial system and production system in Spain at the time of the crisis in a time-explicit disequilibrium-systems model, Figure 4.

The difference between this systems disequilibrium model (Figure 4) and that of the precedingFigure 3 is the intervention of the Spanish Central Bank—as the "lender-of-last-resort" to prevent bank runs in Spanish banks. But this "Bagehot role" of the central bank rescuing the private banks then increased public debt of the Spanish government. Figure 5 compares Spanish public debt to average European public debt. The Spanish government debt was low, compared to the average government debt of EU countries—up until the time of saving the Spanish banks in 2008. The Spanish government debt rose to Eurozone averages.

3. PRIVATE DEBT MARKET—DISEQUILIBRIUM SYSTEMS FINANCIAL MODEL AND ECONOMIC RECESSION IN THE CASE OF SPAIN 2010

Next the fiscal crisis in Spain triggered an economic crisis—a severe recession in the Spanish economy. This was again classical economic system behavior—a fiscal crisis triggered an economic crisis. We next connect the fiscal crisis in Spain to the economic crisis which followed, modeling the connection in a systems dynamics model of a society, as composed of the different subsystems: economic, cultural, political, and technological. Therein, one can display the interaction (between the financial system at the time of a financial bubble and the production system) by showing the models of the financial system upon a topological plane separate from models of the production system on a different model plane [7] . This is depicted in Figure 6, where commodity markets are placed upon a production plane and financial markets are placed upon a financial plane.

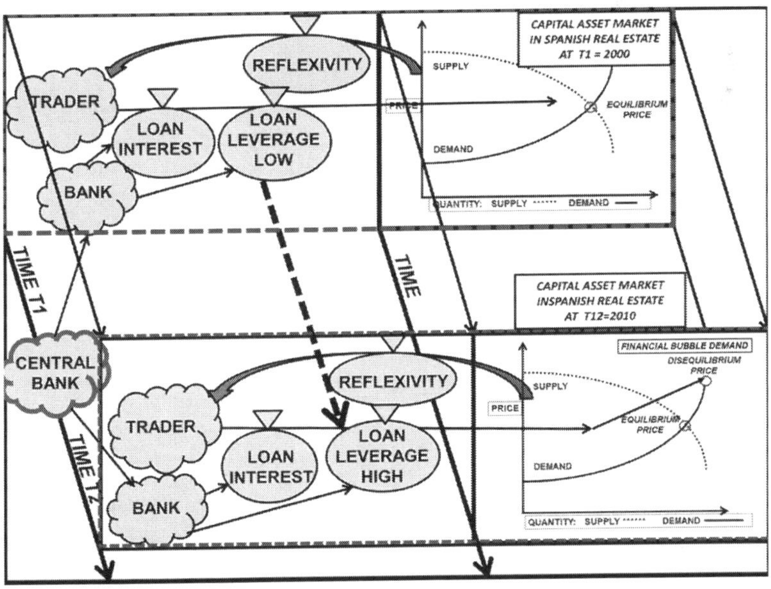

Figure 4. Spanish private debt bubble with Spanish central bank intervention.

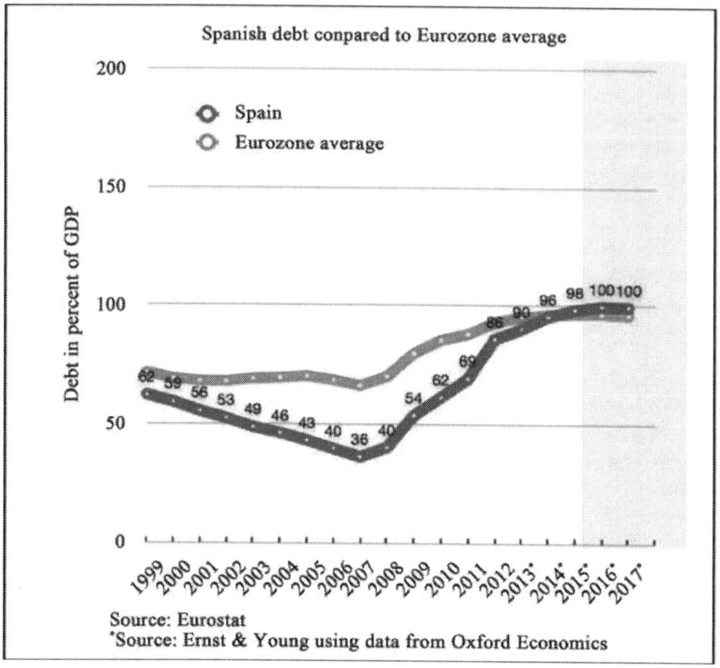

Figure 5. Spanish debt compared to Eurozone average debt source:http://en.wikipedia.org/wiki/European_debt_crisis

In the production sub-system of a national economy, individual businesses, firms, produce goods within the industrial framework of an "industrial value chain". The firm can be modeled as a value-adding system; and the industrial context as an industrial value chain. All the firms and their industrial value chains operate in the production of commodities under an equilibrium pricing of supply-demand. Thus commodity products have a single economic value of "consumption"—which is priced in equilibrium of supply and demand.

In contrast, in the financial sub-system of a national economy, the buying-and-selling (trading) of capital assets need to be modeled in a time-dependent transactional process. Central to the profitability of capital asset trades is the "leverage" in the loans—with higher leverage creating larger profits in the liquidity of an asset (and also higher risk). The time dependence in the financial market allows "leverage" in financial transactions to increase over time with liquidity at higher and higher prices—a financial bubble.

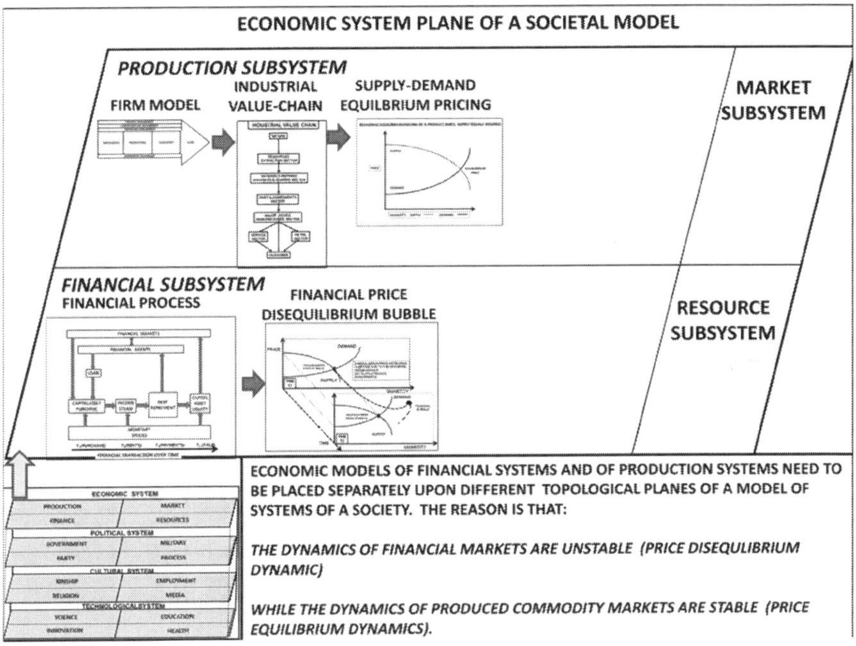

Figure 6. Economic system plane of a societal model.

The reason for topological separation of the models commodity and financial markets is that their dynamics are quite different: price equilibrium in commodity markets but price disequilibrium in financial markets. And a fiscal crisis in the financial system triggers an economic recession in the production plane. This occurs as bank failures stop the credit system of a nation; and business declines with unemployment rising. Banks provide the daily accounts and credit lines upon which businesses operate—manufacturing, services, and retail businesses. When bank accounts are frozen, the monetary system (currency and credit) stops and the economy seizes up. Businesses need daily credit lines with banks to operate on a daily schedule. Thus the monetary system is one of the societal sub-systems (in addition to energy sub-systems) which need to operate on a daily (and even hourly) basis for a modern economy to run continuously.

In Spain in 2011, this connection between financial and production systems was direct and swift. Spain's financial system plane (with its sudden property-bubble/bank-failures) and Spain's production system plane connected (with an immediate Spanish economic-recession). Unemployment in Spain jumped to 36% in 2008 and, even as late as 2012, ran at 24%. This connection is described in the disequilibrium systems model of Figure 7.

In the Financial Sector, a price-disequilibrium systems model depicts the flow of loans from banks to traders (construction businesses and home owners) to develop, sell, and own residences. The reflexivity (market per ception) was that housing prices would rise over time. And prices in the housing market did rise above an equilibrium price because new home construction was continually anticipating expanding demand. Thus in the growing home market, financial expectations (reflexivity) anticipated ever higher prices—driving the market from a price equilibrium to disequilibrium economic condition.

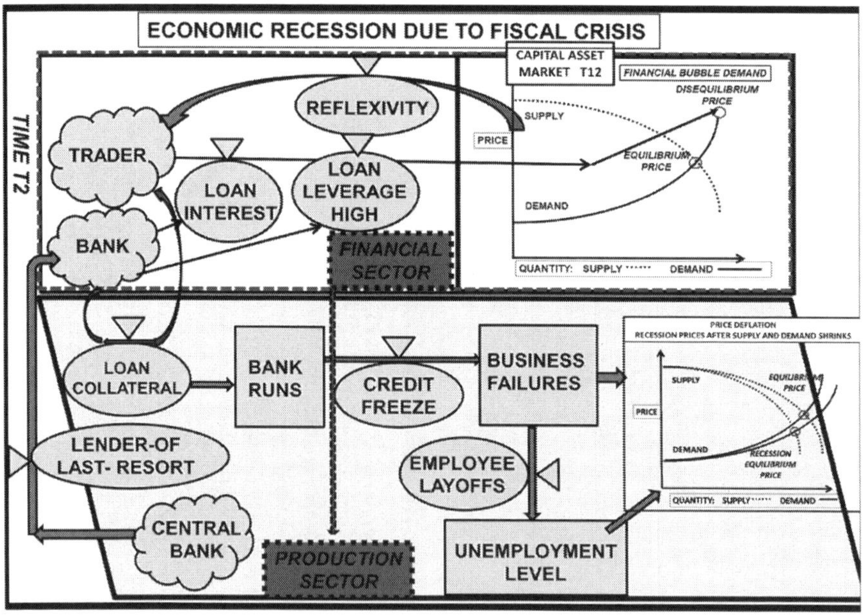

Figure 7. Economic recession due to fiscal crisis.

In the Production Sector, the loan collateral held by the banks in their funding of financial market speculation became worthless (unsellable) with the collapse of the housing market. Instantly, the banks became illiquid (and some insolvent). Bank runs began by depositors withdrawing their accounts. The illiquid/insolvent banks freeze their accounts, and credit ceases in the economy. Businesses and employees are unable to access their bank accounts and credit lines stop working.

In providing construction loans and residence purchase mortgages as collateral the Spanish banks held highly- leveraged loans, with liens on the titles to the properties. When the housing bubble burst, the collateral could not be sold, and the Spanish banks became insolvent. They thus held long-term loans but funded by short-term borrowing from foreign banks and investors. Borrow short and lend long—is a basic way for banks to become insolvent. Thus from illiquid collateral, bank runs occurred on the insolvent banks. Insolvent, the banks could make no further loans and the credit market in Spain froze. Businesses operating daily upon credit lines with banks could no longer obtain cash and closed down production levels, laying off workers. Unemployed labor could not purchase commodities, so that prices in the commodity markets in Spain declined to recession/depression levels. And Spain was in a depression, with high unemployment, lowered production, and a financial crisis.

It was in May 2012 when the recession deepened, and the Spanish government began "bailing-out" banks. David Jolly had reported: "Spanish retail sales fell in April by a record 9.8 percent from a year earlier, the 22nd consecutive monthly decline, the government statistics institute said in Madrid. In a separate report, the Bank of Spain forecast that the economy would continue contracting throughout the first half of 2012, after shrinking by 0.3 percent in the first three months of the year. Spain, with the highest jobless rate in Europe, is struggling to fix its financial sector, which was battered by the collapse of the credit bubble. The country has again become the subject of concern after Bankia, a real estate lender that the government is nationalizing, said Friday that it needed a total bailout of 23.5 billion Euros, or $29.4 billion… The Spanish economy is shrinking as the central government and the regional authorities cut spending to meet budget targets. Making matters worse, the housing sector continues to decline and sales to other European countries, which have helped to sustain exporters, are likely to weaken as the bloc's economy stagnates. The banking sector's troubles could make it harder for even healthy companies to get credit" [8] .

As "lender-of-last-resort", the central bank of Spain moved to "bail out" the failing banks; and then the interest rate on Spanish sovereign bonds jumped higher. In June 2012, David Jolly reported: "Government borrowing costs spiked to euro-era highs for Spain on Thursday, and Italy had to pay high rates again to complete a bond auction, as pressure mounted for a unified European response to the region's financial problems. The yield, or interest rate, on Spain's 10-year bonds reached 7 percent, the level that has set off full international bailouts of some other euro zone members, after Moody's Investors Service downgraded the nation's credit rating late Wednesday… Italy's Prime Minister, Mario Monti, said Thursday that more steps needed to be taken at a European level to protect the euro union from the onslaught of the financial markets. It was an assessment shared by the new French president, François Hollande, who appeared with Mr. Monti at a news conference after they met for two hours in Rome" [9] .

In July 2012, Raphael Minder and Stephen Castle reported: "Struggling to meet euro zone financial targets, Prime Minister Mariano Rajoy of Spain introduced his latest package of tough austerity measures Wednesday, including a rise in the sales tax, reversing his previous stance. The package, Mr. Rajoy's fourth set of budget measures in seven months, is intended to reduce the budget deficit by ?5 billion, or $80 billion, over two and a half years. It follows a decision by European Union finance ministers Tuesday to relax Spain's deficit target for this year to 6.3 percent of gross domestic product, rather than the 5.3 percent target that was set only four months ago. But the finance ministers also said they expected Madrid to continue showing progress on its deficit cutting. The new austerity plan came as Finland's prime minister, Jyrki Katainen, issued a warning Wednesday that the euro's predicament was as perilous as at any time in the past two years. "This situation is dangerous, very dangerous," he said in an interview with Finland's biggest daily, HelsinginSanomat. One of the main elements of the new round of Spanish austerity measures is a rise in the value-added tax to 21 percent from 18 percent" [10] .

In 2012, the austerity pressure on Spanish people was heavy and many exported savings and/or also left Spain. In September 2012, Landon Thomas reported: "After working six years as a senior executive for a multinational payroll-processing company in Barcelona, Spain, Mr. Vildosola is cutting his professional and financial ties with his troubled homeland. He has moved his family to a village near Cambridge, England, where he will take the reins at a small software company, and he has transferred his savings from Spanish banks to British banks. "The macro situation in Spain is getting worse and worse", Mr. Vildosola, 38, said last week just hours before boarding a plane to London with his wife and two small children. "There is just too much risk. Spain is going to be next after Greece, and I just don't want to end up holding devalued pesetas". Mr. Vildosola is among many who worry that Spain's economic tailspin could eventually force the country's withdrawal from the euro and a return to its former currency, the peseta. That dire outcome is still considered a long shot, even if Spain might eventually require a Greek-style bailout. But there is no doubt that many of those in a position to do so are taking their money—and in some cases themselves—out of Spain" [11] .

4. PUBLIC DEBT MARKET—DISEQUILIBRIUM SYSTEMS MODEL IN THE CASE OF GREECE 2010

In the case of Greece, financial contagion was due to government borrowing for funding current operations— covering an operational annual budget deficit. It was simply (and continually) bad fiscal policy. In a previous paper [2] , the disequilibrium systems model for Greece was described and now is shown again in Figure 8.

In this public debt bubble, Greek governments had issued new sovereign bonds each year to fund annual budget deficits. Over time, this continual funding of budget deficits built up the national debt of the country. The Greek national debt as a percentage of GDP (%ND/GDP) rose to about the Eurozone average of 50% ND/GDP in 1985 to 100% in 1993 and then (after 2007) jumped to 180% in the crisis year of 2011. Then Greece needed to annually roll-over 20% (ND/GDP), with its 80% of sovereign debt held by foreigners. This need 1) for annual roll-over mostly to foreigners along with 2) the need to annually continue to add debt to cover budget deficits—spooked the bond market—"market reflexivity". From 2009 to 2012, interest rates on new Greek bonds demanded by the market jumped from 3.7% to 177.37%. The Greek government was bankupt in 2010—despa-rately needing an EU "bail out".

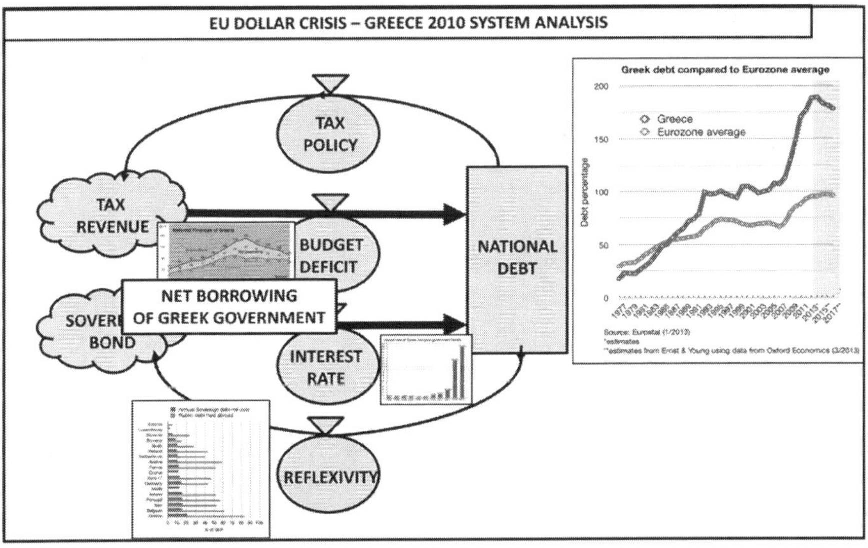

Figure 8. Disequilibrium systems model of Greek fiscal policy

Time is implicit within this systems dynamics models as the flows are continual between sources (e.g. tax sources and bond sources) and stock (e.g. national debt) of the system. Such flows are determined by the institutional structure of the system, adjusted by the structural rate controls (e.g. budget deficit, interest rate, reflexivity, tax policy). If the institutional structure does not alter, then a systems dynamics model is a "stasis" model (constant flow model). Thus the dimension of time did not appear explicitly in this disequilibrium system model of Greek fiscal policy, because there were no structural changes to the Greek system over the period of time leading to the fiscal crisis in 2010. But then 'market reflexivity' (perceptions by bond traders on the likelihood of a Greek debt default) changed! Then bond traders would no longer purchase Greek sovereign bonds at reasonable interest rates. The market would no longer buy them, and interest rates on new Greek bonds jumped—a "Minsky moment". At the time of the fiscal crisis in 2009, the reflexivity in the bond market jumped the acceptable interest rates on new Greek bonds to bubble heights, as market doubt began about the financial stability of the Greek government—a structural change in the market, due to market reflexivity.

Afterwards, the Greek government could no longer sell sovereign bonds in the international bond market and required injections of external funds to continue government operations and a national financial system. A European "bailout" was needed to avoid Greece's default on its debt. In May 2010, a "Troika" (the European Commission, the European Central Bank (ECB) and the International Monetary Fund (IMF)) provided to the Greek Government (then under Prime Minister Georg Papandreou) a first bailout package of 107.3 billion euro. In March 2012, the Troika provided a second bailout package of an additional 130 billion euro for the years 2012-2014. Liz Alderman wrote: "The bailouts came with conditions. Lenders imposed harsh austerity terms, requiring deep budget cuts and steep tax increases. They also required Greece to overhaul its economy by streamlining the government, ending tax evasion and making Greece an easier place to do business" [12] . The "time line" of structure in the Greek fiscal crisis from T1 (2009) to T2 (2015) unfolded as:

T1 "December 2009. Credit ratings agencies downgrade Greece on fears that it could default on its debt". "May 2010. Europe and Greece reach a $146 billion rescue package, conditional on austerity measures. Some economists say the required cuts could kill the patient". "October 2011. Banks agree to take a 50 percent loss on the face value of their Greek debt". "July 2012. Stocks soar after the head of the E.C.B. says policy makers will do ''whatever it takes'' to save the euro zone". T2 "January 2015. Greek voters choose an anti-austerity party. Alexis Tsipras becomes prime minister". "May 2015. Greece quells fears of an imminent default on its debts, authorizing a big loan payment to the IMF It is not clear how much longer Greece can continue to scrape by". June 2015. "Greece elevated the game of brinkmanship with its international lenders on Tuesday (June 2, as Prime Minister Alexis Tsipras announced that his government had sent creditors a new proposal, on Greece's own terms, to unlock financial aid that the struggling nation desperately needs. But the creditors —the European Commission, the International Monetary Fund and the European Central Bank—were moving on Tuesday toward completing their own proposal to break the months long impasse with Athens. That move came after European leaders, including Chancellor Angela Merkel of Germany, held an emergency session in Berlin on Monday night to come up with a blueprint" [13] .

Thus a structural change occurred in the systems model of the Greek fiscal system, with the addition of "bailout" funding from the 'Troika'; and this is shown in Figure 9. One needs to add a central bank source (Greece Central Bank) to receive funds from outside Greece as the "bailout" loans from the Troika. From T1 (2009) to T2 (2015), Greek fiscal policy was on a Troika financial "life-line" (bail-out), unable to handle its debt load— while also having to begin paying back IMF loans.

Next to model the structural change from T1 to T2, we introduce an explicit time dimension (a la "Minsky") into a disequilibrium systems-model for the Greek fiscal crisis—marking changes in Greece's financial structures of its fiscal policy (Figure 10).

This describes a structurally-time-changed-disequilibrium systems model of a governmental fiscal system (public debt). There are two kinds of time lines in a Minsky systems dynamics model of a price-disequilibrium of a financial market. Within the systems dynamics model, the financial processes are in a stasis mode (recurrent mode), with valves (such as tax policy, budget deficit, interest rate, reflexivity) controlling the steady-state flow of finances from sources (tax revenue and sovereign bonds to a financial stock (national debt). But a second kind of time occurs when a bubble builds in the market (from T1 to T2) resulting in a "structural change" in the market system, collapse of the bubble and external banking intervention (Euro Troika) to bail-out the collapsed market.

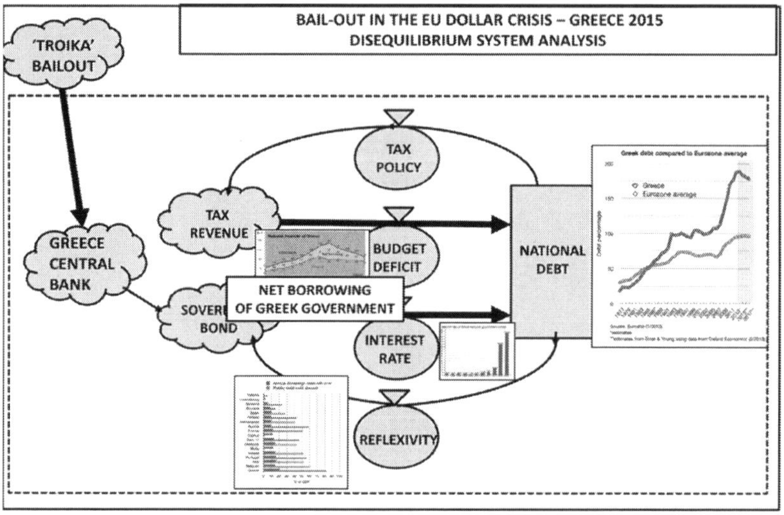

Figure 9. Troika bailout of the Greek central bank in the EU crisis.

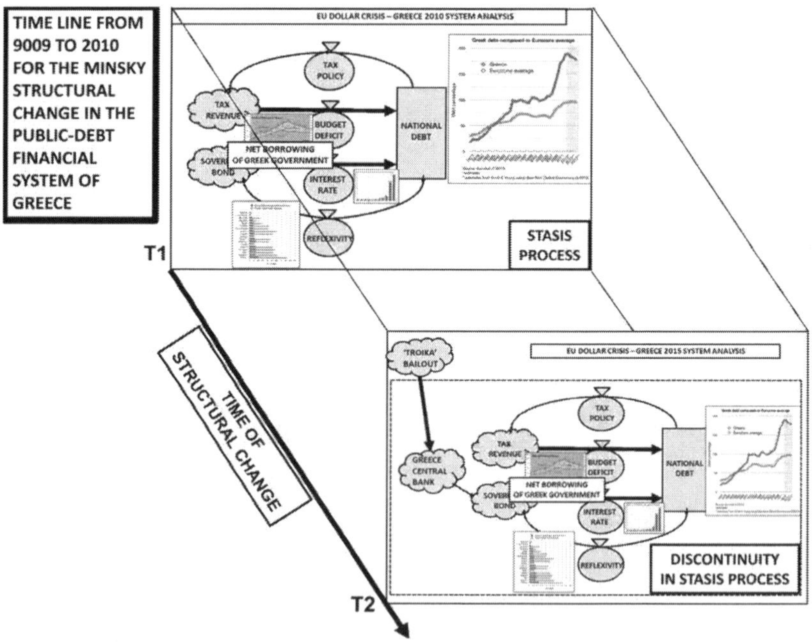

Figure 10. Time line of structural change in the Greek public debt fiscal system.

Financial markets while in stasis (steady-state operation) are inherently not stable over time when structural change occurs (e.g. change in market reflexivity). This is the Keynes/Minsky disequilibrium theory of instability in financial markets over time (time of structural change in the market)—indigenous instability in a financial market.

Structural changes in financial markets occur from either from regulatory changes (government) or from reflexivity changes (trader perceptions) about how the trading operations occur in markets of capital assets—Keynes/Minsky financial markets.

Models of national financial systems need to be embedded into an international financial system, when structural changes occur in national systems.

5. RESULTS

In modeling different countries' crises in the Eurozone, one observes different debt markets triggering the financial bubbles—either as public debt (sovereign bond markets) or as private debt (e.g. residential markets). Greece was a pure economic case of a public debt bubble and Spain was a pure economic case of a private debt bubble. All the other countries impacted by the Euro crisis (such as Ireland, Portugal, Italy) were cases of one or the other or a mixture of both. The importance of distinguishing the market source of a disequilibrium price crisis is that different policies are needed to solve a crisis problem, to "fix" a crisis.

In the case of Spain (with a private debt crisis) the economy could be fixed by preventing bank runs and regulating the bank system—recapitalizing the illiquid banks, and closing down the insolvent banks. The Spanish central bank was able to accomplish this with the 'bail-out' of European Union loans to the Spanish government. And Spain began to emerge from the 2010 crisis three years later in 2013. Raphael Minder reported: "Spain pulled out of a two-year recession in the third quarter (October 2013), with its economy growing 0.1 percent from the previous quarter, according to preliminary data released Wednesday by the Spanish central bank... The growth news will likely provide a fillip for the conservative government of Prime Minister Mariano Rajoy, which, since coming into office almost two years ago, has pushed through deeply unpopular spending cuts and tax hikes in order to comply with budgetary targets in accordance with Spain's European partners... In the end, Spain used 41 billion Euros, or $56 billion, of the 100 billion Euros that it negotiated in a European banking rescue package" [14] .

In contrast in the case of Greece (with a public debt crisis), the Greek economy continued in crisis and depression from 2010 into 2015—still with no "fix" (despite sizable loans from the European Union). In 2015, Hugo Dixon wrote: "The debt is still huge—313 billion Euros, or about $353 billion—175 percent of the Greek gross domestic product. But the bulk, about ?84 billion, is now owed to Eurozone governments and pays low interest. What is more, Greece doesn't have to start repaying it until 2020 and then has over 30 years to finish. The so-called present value of this debt, what it is worth today, is therefore much lower than its official face value. Unfortunately, this isn't the full story. Greece has four other sources of borrowing: ?7 billion owed to the European Central Bank; ?0 billion owed to the International Monetary Fund; debt owed to private investors; and short-term treasury bills. The private debt isn't an immediate problem, because most of it has also been stretched out over a long period. Nor are the treasury bills, as Athens is still able to roll them over as they come due. But the ECB and IMF debt is problematic because most of it comes due in the next five years. In fact, there is a crunch in the next few months, with over ?0 billion in repayments due by the end of September. The snag is that IMF and ECB debt cannot be rescheduled without breaching rules and treaties. This is why the focus of the current talks is about lending Greece more money, so that it can meet these obligations... The Eurozone will also be under pressure to provide yet more loans to Athens because the money left in the current bailout program won't be sufficient to cover all the repayments due to the E.C.B. and I.M.F. What is more, the country has other needs, such as paying overdue bills to suppliers, recapitalizing Greek banks and repaying the loans it has forcibly extracted from local authorities within Greece" [15] .

6. CONCLUSIONS

The Eurodollar crises of 2009-2010 provide empirical evidence about policies for dealing with two different debt markets, private and public. Disequilibrium-systems models of how private and public financial markets can move toward price disequilibrium are useful for identifying 1) how fiscal policies can create fiscal crises and 2) how central bank policies can help or not help to solve fiscal crises.

What was particularly interesting in this comparison was that the European Union (including its European Central Bank and the IMF) was able to help solve the Spanish financial crisis but not the Greek financial crisis. This difference in "crisis solvability" depended upon the underlying fiscal models of private debt markets compared to the public debt markets. Accordingly, the two disequilibrium-systems models of financial markets in the empirical case of the Eurodollar crises suggest the following hypothesis: Central banks can solve crisis- problems of private debt but not crises of public debt. A central bank has no control over its government's tax policy.

REFERENCES

1. Betz, F. (2014) Disequilibrium Pricing Theory. Theoretical Economics Letters, 4, 60-67.
2. Betz, F. (2014) Disequilibrium Pricing—Greek Euro Crisis. Theoretical Economics Letters, 4, 897-909.
3. Minsky, H. (1975) John Maynard Keynes. Columbia University Press (Reprinted in 2008 by McGraw Hill Education), New York City.
4. Wolf, M. (2014) The Shifts and the Shocks: What We've Learned-and Have Still to Learn-from the Financial Crisis. Penguin Press, New York City.
5. Burridge, T. (2012) Spanish Banks to Get up to 100bn Euros in Rescue Loans. BBCNEWS, June 12. http://www.bbc.com/news/business-18382659
6. Norris, F. (2011) Spain's Banking Mess. New York Times, September 22.http://www.nytimes.com/2011/09/23/ business/spains-banking-mess.html
7. Betz, F. (2013) Why Bank Panics Matter. Springer, New York.
8. Jolly, D. (2012) Spanish Data Point to Further Decline. New York Times, May 29.http://www.nytimes.com/2012/05/30/business/global/daily-euro-zone-watch.html?_r=0
9. Jolly, D. (2012) Spanish Cost of Borrowing Soars After Rating Cut. New York Times, June 14. http://www.nytimes.com/2012/06/15/business/global/daily-euro-zone-watch.html
10. Minder, R. and Stephen, C. (2012) Spain Plans New Round of Tough Austerity Measures. New York Times, July 11.http://www.nytimes.com/2012/07/12/business/global/daily-euro-zone-watch.html
11. Thomas, L. (2012) Fears Rising Spaniards Pull out Their Cash and Get out of Spain. New York Times, September 3.http://www.nytimes.com/2012/09/04/business/global/money-and-people-leave-spain-as-economic-gloom-deepens.html

12. Alderman, L. (2015) Explaining the Greek Debt Crisis. New York Times, April 8.http://www.nytimes.com/2015/04 /09/business /international/explaining-the-greek-debt-crisis.html?_r=0

13. Alderman, L. (2015) Greece Challenges Creditors with New Proposal to Break Debt Impasse. New York Times, June 2.http://www.nytimes.com/2015/06/03/business/international/ greece-debt-ecb-imf-european-commission.html?_r=0

14. Minder, R. (2013) Spain Emerges From Two Year Recession. New York Times, October 23. http://www.nytimes.com/2013 /10/24/business/international/spain-emerges-from-two-year-recession.html

15. Dixon, H. (2015) Financial Engineering to Athen's Rescue. International New York Times, June 14. http://www. nytimes.com/2015/06/15/business/international/financial-engineering-to-athenss-rescue.html?_r=0

8

CHAPTER

Applying the Barycentric Jacobi Spectral Method to Price Options with Transaction Costs in a Fractional Black-Scholes Framework

B. F. Nteumagné, E. Pindza, E. Maré

Department of Mathematics and Applied Mathematics, University of Pretoria, Pretoria, Republic of South Africa

ABSTRACT

The aim of this paper is to show how options with transaction costs under fractional, mixed Brownian-fractional, and subdiffusive fractional Black-Scholes models can be efficiently computed by using the barycentric Jacobi spectral method. The reliability of the barycentric Jacobi spectral method for space (asset) direction discretization is demonstrated by solving partial differential equations (PDEs) arising from pricing European options with transaction costs under these models. The discretization of these PDEs in time relies on the implicit Runge-Kutta Radau IIA method. We conducted various numerical experiments and compared our numerical results with existing analytical solutions. It was found that the proposed method is efficient, highly accurate and reliable, and is an alternative to some existing numerical methods for pricing financial options.

Keywords: Jacobi Spectral Method; European Options; Fractional Black-Scholes Model; Mixed Brownian-Fractional Brownian Black-Scholes Model; Transaction Cost; Subdiffusive Fractional Black-Scholes Model; Scaling

1. INTRODUCTION

Modeling of financial derivatives has been of great interest in the past three decades. Numerous mathematical models have been developed from the classical Black-Scholes [1] framework to help investors in their decision making process. These models are based on an arbitrage argument, i.e., by continuously adjusting a portfolio consisting of a stock and a risk-free bond, an investor can exactly replicate the return to any option on the stock.

In the presence of transaction costs in capital markets, the presence of an arbitrage argument [2] is no longer valid [3], since perfect hedging is impossible. Due to the infinite variation of the geometric Brownian motion, the continuous replication policy incurs an infinite amount of transaction costs over any trading interval no matter how small it might be. Therefore, in recent years, one observes many generalizations of the Black-Scholes model to deal with the problem of option pricing and hedging with transaction costs. This leads to the Black-Scholes model but with an adjusted volatility. Leland [3] was the first to examine option replication in a discrete time setting, and proposed a modified replicating strategy, which depended upon the level of transaction costs, the revision interval, the option to be replicated and the environment. Subsequently, several authors proposed new models, but all in discrete time [4,5]. However, these models based on the diffusion process known as geometric Brownian motion (GBM) have severe shortcomings, for example, long-range correlations, heavy-tailed and skewed marginal distributions, lack of scale invariance and periods of constant values, to enumerate these only. Recently, Wang[6] obtained a European call option pricing formula using a mean self-financing delta hedging argument in a discrete time setting and then showed how scaling and long-range dependence impacted dramatically on the pricing of options using the fractional and multifractional Black-Scholes model with transaction costs.

In continuous time, a lot of efforts have been done in order to alleviate the problem of an infinite amount of transaction costs over any trading interval when the asset process follows a geometric Brownian motion. Magdziarz [7] et al. introduced a subdiffusive geometric Brownian process which captures the subdiffusive characteristics of financial markets. They showed that their model is arbitrage-free. The same idea was used later by Wang et al. [8], Hui and Yun-Xiu [9] to obtain a Black-Scholes equation with transaction costs in subdiffusive fractional Brownian motion regime. However, closed-form solutions of these PDEs in finance are generally rare. Therefore, one has to consider numerical methods to obtain solutions.

In this paper, we are concerned with the application of the barycentric Jacobi interpolation to value options with transaction costs under fractional [6], mixed Brownian-fractional [10] and subdiffusive fractional [8] Black-Scholes models. Barycentric spectral methods were introduced by Baltensperger et al. [11] to solve boundary value problems. Recently, these methods have emerged in the field of finance as a promising tool to solve option pricing problems. Pindza and Patidar [12] proposed an accurate method, namely the barycentric Chebyshev spectral method, to price options in illiquid markets. Ngounda et al. [13] used the barycentric Chebyshev domain decomposition method to provide fast and accurate results for pricing European options with jumps, which was later extended and applied to Heston's volatility model (see [14]). Most of the work on barycentric spectral methods has been based on the use of uniform or Chebyshev grids. Recently, Wang et al. [15] computed explicit barycentric weights for Jacobi polynomial interpolation in the roots or extrema of classical orthogonal polynomials in terms of the nodes and weights of the corresponding Gaussian quadrature rule. Hence, we investigate the utility of this new barycentric interpolation in the field of finance. The semi-discretization of the PDE in time is realized by using a 7-stage 13th-order fully implicit Runge-Kutta Radau IIA method with adaptive time stepping [16].

This paper is structured as follows. Section 2 reviews the option pricing formulation with transaction costs under fractional, mixed Brownian-fractional and subdiffusive fractional processes. In Section 3, we introduce the Jacobi barycentric spectral method, semi-discretize the PDE in the asset space and propose a conformal map in order to improve the accuracy of our method. In Section 4, we perform numerous experiments in order to advocate the utility of the barycentric spectral method. Finally, Section 5 gives a summary and scope for further research.

2. PRICING WITH TRANSACTION COSTS UNDER FRACTIONAL, MIXED BROWNIAN-FRACTIONAL AND SUBDIFFUSIVE FRACTIONAL PROCESSES

Let $\mathrm{M} = \left(\Omega, \mathrm{F}, \mathrm{P}, (\mathrm{F}_t)_{t\geq 0}, (S_t)_{t\geq 0}\right)$ be a complete probability space carrying a fractional Brownian motion $(B_H(t))_{t\in\mathbb{R}}$ with Hurst exponent $H \in (0,1)$, where Ω is the set of all possible outcomes of the experiment known as the sample space, $\mathbf{F}$ is the set of all events, $\mathbf{P}$ is a real world probality, F_t is a natural filtration, S_t a risky underlying asset price process. Assume that the price S_t of the underlying stock at time t satisfies a fractional Black-Scholes model

$$S_t = S_0 \exp\left(\mu t + \sigma B_H(t)\right), \tag{1}$$

where μ, $H \in (0,1)$ and $S_0 > 0$ are constants. Assume that the portfolio is revised every small fixed time step δt; transaction costs are proportional to the value of the transaction in the underlying. Let k denote the round trip transaction cost per unit dollar of transaction. Then under all the assumptions given by Wang [6], the Black-Scholes equation with transaction costs assuming factional Brownian motion can be written as

$$\frac{\partial u}{\partial t} + rS\frac{\partial u}{\partial S} + \frac{\tilde{\sigma}^2}{2}S^2\frac{\partial^2 u}{\partial S^2} - ru = 0, \tag{2}$$

where the modified volatility is given by

$$\tilde{\sigma}^2 = \sigma^2\left[(\delta t)^{2H-1} + Le(H)\,\mathrm{sign}(\Gamma)\right]. \tag{3}$$

Here $Le(H)$ is the fractional Leland number [6].

If we assume that the price S_t of the underlying stock at time t satisfies a mixed Brownian-fractional Brownian model

$$S_t = S_0 \exp\left(\mu t + \sigma B(t) + \sigma_H B_H(t)\right), \tag{4}$$

then under all the assumptions given by Wang [10], the Black-Scholes equation with transaction costs in mixed Brownian-fractional Brownian motion can be written as

$$\frac{\partial u}{\partial t}+rS\frac{\partial u}{\partial S}+\frac{\tilde{\sigma}^2}{2}S^2\frac{\partial^2 u}{\partial S^2}-ru=0, \tag{5}$$

where the modified volatility is given by

$$\tilde{\sigma}=\left(\sigma^2+\sigma_H^2\left(\delta t\right)^{2H-1}+k\sqrt{\frac{2}{\pi}\left(\frac{\sigma^2}{\delta t}+\sigma_H^2\left(\delta t\right)^{2H-2}\right)}\operatorname{sign}\left(\Gamma\right)\right)^{\frac{1}{2}}, \tag{6}$$

With

$$\sigma_H^2=\frac{\frac{2k^2}{\pi}+\sqrt{\left(\frac{2k^2}{\pi}\right)^2+\frac{8k^2}{\pi}\sigma^2\delta t}}{2},$$

$$\delta t=\left(\frac{2}{\pi}\right)^{\frac{1}{2H}}\left(\frac{k}{\sigma}\right)^{\frac{1}{H}}\text{ and }\Gamma=\frac{\partial^2 u}{\partial S^2}.$$

A subdiffusive fractional Brownian motion is described by

$$\mathrm{d}S(t)=\mu S(t)(\mathrm{d}t)^{2H}+\sigma S(t)\mathrm{d}B(t). \tag{7}$$

Wang et al. [8] obtained the modified volatility corresponding to the continuous Black-Scholes equation with transaction cost in subdiffusive regime as

$$\frac{\partial u}{\partial t}+rS\frac{\partial u}{\partial S}+\frac{\tilde{\sigma}_H^2}{2}S^2\frac{\partial^2 u}{\partial S^2}-ru=0, \tag{8}$$

where the modified volatility is given by

$$\tilde{\sigma}_H^2=2Ht^{2H-1}\left(\frac{t^{\alpha-1}}{\overline{\Gamma}(\alpha)}\right)^{2H}\left(1+\frac{k}{2}\left|\frac{2\mu}{\sigma^2}-1\right|\operatorname{sign}(\Gamma)\right)\sigma^2. \tag{9}$$

Here $\overline{\Gamma}(\alpha)$ represents the gamma function evaluated at α and $\operatorname{sign}(\Gamma)$ is the second derivative of the option value with respect to the asset.

The difference between American and European call and put options is made by the initial and boundary conditions. In this work we focus exclusively on the European call option. Such options have the following initial and boundary conditions

$$\begin{aligned} &u(0,t)=0, \quad u(S,0)=\max\{S-K,0\},\\ &u(S,t)=S-Ke^{-rt}, \text{ as } S\to\infty, \end{aligned} \tag{10}$$

where $S\in[0,\infty]$ and $t\in[0,T]$. We want to solve the three PDEs (2), (5) and (8) subjected to modified volatilities (3), (6) and (9), respectively. Before moving to the applications of the barycentric Jacobi spectral method to solve these problems, it is worthwhile to consider some preliminaries of this method.

3. BARYCENTRIC JACOBI SPECTRAL COLLOCATION METHOD

In this section, we turn our attention to the problem of barycentric Jacobi interpolation and present a fast algorithm for the efficient computation of the interpolation formula using Jacobi-Gauss-Lobatto (JGL) points. This interpolation is realized by a class of the Lagrange form of the interpolating polynomial, as follows. Let x_j, $j = 0,1,\cdots,N$ be a set of distinct nodes. Then the polynomial of degree N that interpolates the function $u(x)$ at these points is given by[17]

$$p_N(x) = \sum_{j=0}^{N} u_j \ell_j(x),\ \ell_j(x) = \prod_{k=0,1,k\neq j}^{N} \frac{x - x_k}{x_j - x_k}, \tag{11}$$

where the Lagrange polynomial ℓ_j corresponding to the node x_j has the property

$$\ell_j(x_k) = \begin{cases} 1, & j = k \\ 0, & \text{otherwise}, \end{cases} \tag{12}$$

with $j,k = 0,1,\cdots,N.$ Generally, the Lagrange form of the interpolating polynomial (11) is not advocated for numerical computations. In particular, it requires $O(N^2)$ additions and multiplications for each evaluation of $p_N(x)$ and every time a node x_j is modified or added, all Lagrange fundamental polynomials have to be recalculated. However, with slight modifications, the Lagrange formula is indeed of great practical use. This has been noted by several authors, including Henrici [18] and Werner [19]. Berrut and Trefethen [20] modified the Lagrange polynomial through barycentric interpolation and proposed an improved Lagrange formula.

Following [20], we define $\ell(x)$, the numerator of ℓ_j in (11) divided by $x-x_j$, as

$$\ell(x)=\prod_{j=0}^{N}(x-x_j). \tag{13}$$

In addition, if we define the barycentric weight by

$$\lambda_j=\frac{1}{\prod_{k=0,k\neq j}^{N}(x_j-x_k)},\quad j=0,1,\cdots,N, \tag{14}$$

i.e., $\lambda_j=1/\ell'(x_j)$, then ℓ_j in (11) becomes

$$\ell_j(x)=\ell(x)\frac{\lambda_j}{x-x_j}. \tag{15}$$

Consequently, the Lagrange formula (11) becomes

$$p_N(x)=\ell(x)\sum_{j=0}^{N}\frac{\lambda_j}{x-x_j}u_j. \tag{16}$$

In particular, if $u(x)=1$, we obtain

$$1=\sum_{j=0}^{N}\ell_j(x)=\ell(x)\sum_{j=0}^{N}\frac{\lambda_j}{x-x_j}. \tag{17}$$

Dividing (16) by (17), we get the barycentric formula for p_N as

$$p_N(x)=\frac{\sum_{j=0}^{N}\frac{\lambda_j}{x-x_j}u_j}{\sum_{j=0}^{N}\frac{\lambda_j}{x-x_j}}. \tag{18}$$

This is the most used form of Lagrange interpolation in practice and admits $O(N)$ operations. In order to obtain good approximations via interpolation, the choice of interpolation nodes and barycentric weights is particularly important. For certain particular sets of points, such as equidistant points as well as Chebyshev points, the barycentric weights λ_j can be computed analytically[20]. Recently, Wang et al. [15] computed explicit barycentric weights for polynomial interpolation in the roots or extrema of classical orthogonal polynomials in terms of the nodes and weights of the corresponding Gaussian quadrature rule.

In this paper, we are interested in the barycentric Jacobi interpolation. The Jacobi-Gauss-Lobatto quadrature rule is defined by

$$\int_{-1}^{1}u(x)=\sum_{j=0}^{N}w_j u(x_j), \tag{19}$$

where the Jacobi-Gauss-Lobatto points $\{x_j\}_{j=0}^{N}$ are the zeros of $(1-x^2)P_{N-1}^{(\alpha,\beta)}(x)$ and $P_N^{(\alpha,\beta)}(x)$ is the Jacobi polynomial of degree N. The following result gives an analytical formula of barycentric weights for the Jacobi-Gauss-Lobatto points.

Theorem 3.1 ([15]) Let $x_j, j=0,\cdots,N$, be the roots of $(1-x^2)P_{N-1}^{(\alpha,\beta)}(x)$ and let w_j be the corresponding weights of the interpolatory quadrature rule with weight function $(1-x)^{\alpha}(1+x)^{\beta}$. Then the simplified barycentric weights are for Jacobi-Gauss-Lobatto points, the simplified barycentric weights λ_j are given by

$$\lambda_j = (-1)^j \sqrt{\delta_j w_j},$$
$$\delta_j = \begin{cases} \beta+1, & j=0 \\ \alpha+1, & j=N \\ 1, & \text{otherwise}, \end{cases} \tag{20}$$

The proof of Theorem 3.1 can be found in [15].

The computation of entries of the first and second order differentiation matrices $D^{(1)}$ and $D^{(2)}$ where

$$D_{ij}^{(1)} = \ell'_j(x_i) \text{ and } D_{ij}^{(2)} = \ell''_j(x_i), \tag{21}$$

is given as in [11] by

$$\ell'_j(x_i) = \begin{cases} \dfrac{\lambda_j/\lambda_i}{x_i - x_j}, & \text{if } i \neq j, \\ -\sum_{i \neq j}^{N} \ell'_j(x_i), & \text{if } i = j, \end{cases} \tag{22}$$

$$\ell''_j(x_i) = \begin{cases} -2\dfrac{\lambda_j/\lambda_i}{x_i - x_j}\left(\displaystyle\sum_{k=0,k\neq i}^{N} \frac{\lambda_k/\lambda_i}{x_i - x_k} - \frac{1}{x_i - x_j}\right), & \text{if } f \neq j, \\ -\displaystyle\sum_{i\neq j}^{N} \ell''_j(x_i), & \text{if } f = j, \end{cases} \tag{23}$$

where $i, j = 0, 1, \cdots, N$.

3.1. Conformal Mappings for High Resolution of Non-Smooth Initial Conditions

It is well-known that the solution of Equation (8) is very sensitive to localization errors when S is in the vicinity of K, because the second derivative of the payoff does not exist at this point. Therefore, to increase accuracy it would be reasonable to use an adaptive mesh with high concentration of the mesh points around $S = K$, while a rarefied mesh could be used far away from this area. If we assume that $S \in [a, b]$, where both a and b are finite, then without further loss of generality we may assume that the interval of integration is $[-1, 1]$, as the linear transformation

$$2S = (b - a)x - (a + b). \tag{24}$$

In this paper, we use the conformal map g given in [21] by

$$x = g(y) = \tilde{\beta} + \frac{1}{\tilde{\alpha}}\sinh\left[\bar{\lambda}(y - \mu)\right], \tag{25}$$

Where

$$\bar{\lambda} = \frac{\gamma+\delta}{2}, \ \mu = \frac{\gamma-\delta}{\gamma+\delta}, \tag{26}$$

with

$$\gamma = \sinh^{-1}\left[\tilde{\alpha}\left(1+\tilde{\beta}\right)\right], \delta = \sinh^{-1}\left[\tilde{\alpha}\left(1-\tilde{\beta}\right)\right], \tag{27}$$

where $\tilde{\alpha}$ and $\tilde{\beta}$ determine the magnitude of the region of rapid change and the location, respectively. Here, y represents the Jacobi-Gauss-Lobatto points.

In Figure 1, we plot the new grid obtained from the original Jacobi y grid by using transformations (25) and (24) together. The new grid in $S(x)$ contains 100 nodes distributed from $a=0$ to $b=250$. Value of parameters used in this example are: $\tilde{\alpha}=10^4$ and $\tilde{\beta}=\left(2K-(b-a)\right)/(b-a)$, where K is the strike price. In addition, we show the transformation function around K for $\tilde{\alpha}=10$ (black), $\tilde{\alpha}=10^2$ (green), $\tilde{\alpha}=10^3$ (purple) and $\tilde{\alpha}=10^4$ (red). We observe that when $\tilde{\alpha}$ decreases the grid approaches a Jacobi-Gauss-Lobatto grid distribution one. When $\tilde{\alpha}$ increases we accommodate more grid points around the strike price.

A significant advantage of the barycentric Jacobi spectral method is that it eliminates tedious computations of transformed derivatives using the chain rule as is usually the case in other spectral collocation methods.

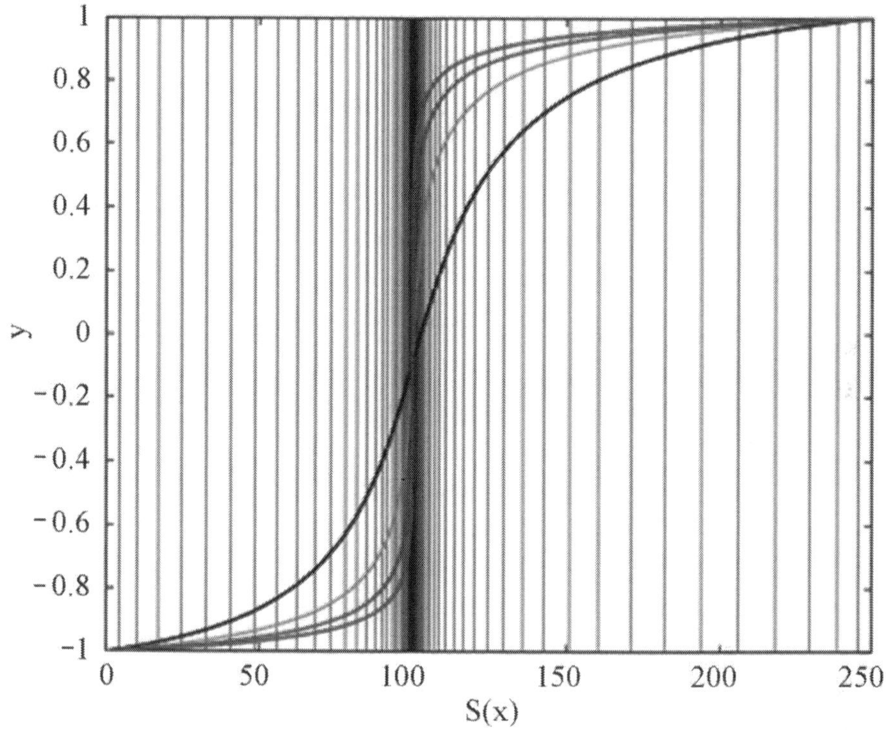

Figure 1. Asset direction grids S obtained from the mapped Jacobi grids x.

3.2. Application to the Black-Scholes PDE

We discretize the Black-Scholes PDEs (2), (5) and (8) in the asset (space) direction by means of barycentric Jacobi spectral method. Let $x = g(y)$ be the transformed Jacobi-Gauss-Lobatto points, the first step is to transform $x \in [-1,1]$ into $S \in [0, S_{\max}]$ that better suits the option at hand using $x = (2S - S_{\max})/S_{\max}$. Now writing $u(S,t) = u(x,t)$, the PDE (8) together with its initial and boundary conditions yields

$$\begin{cases} u_t = p(x)u_{xx} + q(x)u_x + ru, \\ u(x,0) = u^0, \ -1 \le x \le 1, \ 0 \le S \le S_{\max}, \\ u(-1,t) = u_0, \ u(1,t) = u_N, \ 0 \le t \le T, \end{cases} \tag{28}$$

Where

$$p(x)=\frac{1}{2}\sigma^2 S^2\left(\frac{2}{S_{\max}}\right)^2,\ q(x)=(r-\delta)S\left(\frac{2}{S_{\max}}\right).$$

Substituting (11) into (28) yields the following system of nonlinear ODEs

$$\begin{cases} u_t(x,t)=p(x)\sum_{k=0}^{N}u_k(t)\ell_k''(x)+q(x)\sum_{j=0}^{N}u_k(t)\ell_k'(x) \\ \qquad +r\sum_{k=0}^{N}u_k(t)\ell_k(x), \\ u_0=u(-1,t),\ u_N=u(1,t). \end{cases} \tag{29}$$

In order to write (29) in matrix form, we introduce the following matrix and vector notations

$$\begin{cases} u=[u_1,u_2\cdots,u_{N-1}]^{\mathrm{T}}, \\ D^{(1)}=\left(D_{ij}^{(1)}\right),\ D_{ij}^{(1)}=\ell_j'(x_i);\ i,j=1,\cdots,N-1, \\ D^{(2)}=\left(D_{ij}^{(2)}\right),\ D_{ij}^{(2)}=\ell_j''(x_i);\ i,j=1,\cdots,N-1, \\ P=\mathrm{diag}(p(x_i)),\ Q=\mathrm{diag}(q(x_i));\ i=1,\cdots,N-1, \end{cases} \tag{30}$$

where I is an $(N-1)\times(N-1)$ identity matrix. Consequently, (29) can be expressed as an initial value problem of the form

$$\frac{\mathrm{d}u}{\mathrm{d}t}=g(t,u),\ u(0)=u_0. \tag{31}$$

Where

$$g(\boldsymbol{u},t)=A\boldsymbol{u}+\Big[QD^{(1)}\boldsymbol{u}-rI\boldsymbol{u}+\big(p(x_i)D_{i0}^{(2)}+q(x_i)D_{i0}^{(1)}+rI_{i0}\big)u_0$$
$$+\big(p(x_i)D_{iN}^{(2)}+q(x_i)D_{iN}^{(1)}+rI_{iN}\big)u_N\Big]^{\mathrm{T}}. \tag{32}$$

We use a 7-stage 13th-order fully implicit Runge-Kutta Radau IIA method with step size control to integrate the system of ODEs (31). The method is B-stable and stiffly accurate. Details of the method can be found in [16].

4. NUMERICAL RESULTS AND DISCUSSIONS

We apply the barycentric Jacobi spectral method to value Black-Scholes equations with transaction costs under fractional (FBS), mixed Brownian-fractional (MBS) and subdiffusive fractional (SBS) processes. To show the efficiency of the present method we report the root mean square norm error (L_2) of the solution computed with N grid points, given by

$$L_2=\sqrt{\sum_{j=1}^{N}\left(u(S_j)-u_{exact}(S_j)\right)^2}, \tag{33}$$

and the maximal norm error (L_∞)

$$L_\infty=\max_{j=1,\cdots,N}\left|u(S_j)-u_{exact}(S_j)\right|, \tag{34}$$

where $u(S_j)$ and $u_{exact}(S_j)$ are the benchmark and computed values of the solution $\boldsymbol{u}$ at S_j.

The analytic solutions of Black-Scholes equations with transaction costs under FBS, MBS and SBS regime are possible if we assume that the Greek Γ is always positive for the above mentioned models as in the case of standard Black-Scholes model, then for European call options under SBS[6] regime is known, and expressed as

$$V_{SBS}(S,\tau) = SN\left(\hat{d}_1(\tau)\right) - Ke^{-r\tau} N\left(\hat{d}_2(\tau)\right), \tag{35}$$

Where

$$\hat{d}_1(\tau) = \frac{\ln(S/E) + \left(r + \frac{1}{2}\hat{\sigma}^2_{H,\alpha}(\tau)\right)\tau}{\hat{\sigma}_{H,\alpha}(\tau)\sqrt{\tau}},$$
$$\hat{d}_2(\tau) = \frac{\ln(S/E) + \left(r - \frac{1}{2}\hat{\sigma}^2_{H,\alpha}(\tau)\right)\tau}{\hat{\sigma}_{H,\alpha}(\tau)\sqrt{\tau}}; \tag{36}$$

here $\hat{\sigma}^2_{H,\alpha}(\tau) = \frac{\int_\tau^T \tilde{\sigma}^2_{H,\alpha}(t)\,dt}{\tau}$ is given by

$$\hat{\sigma}^2_{H,\alpha} = \frac{\sigma^2}{\alpha(\Gamma(\alpha))^2}\left(1 + \frac{k}{2}\left|\frac{2\mu}{\sigma^2} - 1\right|\right)\left(\frac{T^{2H\alpha} - (T-\tau)^{2H\alpha}}{\tau}\right), \tag{37}$$

$\tau = T - t$ and $N(\cdot)$ is the cumulative probability distribution function for a standardized normal variable

$$N(y)=\frac{1}{\sqrt{2\pi}}\int_{-\infty}^{y}\mathrm{e}^{-\frac{x^2}{2}}\mathrm{d}x. \tag{38}$$

If we assume the same sign behaviour for Γ in the MBS model, then the analytic solution is given by

$$V_{MBS}(S,\tau)=S\mathrm{e}^{-\delta\tau}N(d_1)-E\mathrm{e}^{-r\tau}N(d_2), \tag{39}$$

Where

$$d_1=\frac{\ln(S/K)+\left(r+\frac{\tilde{\sigma}_\tau^2}{2}\right)\tau}{\tilde{\sigma}_\tau\sqrt{\tau}},\ d_2=d_1-\tilde{\sigma}_\tau\sqrt{\tau}, \tag{40}$$

with the modified volatility given by

$$\tilde{\sigma}=\left(\sigma^2+\sigma_H^2(\delta t)^{2H-1}+k\sqrt{\frac{2}{\pi}\left(\frac{\sigma^2}{\delta t}+\sigma_H^2(\delta t)^{2H-2}\right)}\right)^{\frac{1}{2}}. \tag{41}$$

Here $\sigma_H^2=\dfrac{\frac{2k^2}{\pi}+\sqrt{\left(\frac{2k^2}{\pi}\right)^2+\frac{8k^2}{\pi}\sigma^2\delta t}}{2}$ and $\delta t=\left(\frac{2}{\pi}\right)^{\frac{1}{2H}}\left(\frac{k}{\sigma}\right)^{\frac{1}{H}}$. In the case of FBS model, the closed-form solution is given as

$$V_{FBS}(S,\tau) = Se^{-\delta\tau}N(d_1) - Ee^{-r\tau}N(d_2), \tag{42}$$

where

$$d_1 = \frac{\ln(S/K) + \left(r + \frac{\tilde{\sigma}^2}{2}\right)\tau}{\tilde{\sigma}\sqrt{\tau}}, \; d_2 = d_1 - \tilde{\sigma}\sqrt{\tau}, \tag{43}$$

With

$$\tilde{\sigma}^2 = \sigma^2\left[(\delta t)^{2H-1} + Le(H)\right]. \tag{44}$$

In order to test the accuracy of the method, we present a comparison against the above exact solutions. In Figure 2 (top left) we plot together numerical and exact solutions for comparison purposes. We select numerical values of the parameters to be $K = 100$, $\sigma = 0.2$, $r = 0.03$, $T = 1$, $N = 200$, $S_{min} = 0$, $S_{max} = 250$, $H = 0.6$, $\mu = 0.2$, $k = 1$, $\alpha = 0.7$. We choose the tolerance of the 7-stage 13th-order fully implicit RungeKutta Radau IIA method [16] to be $tol = 10^{-10}$, so that the error is dominated by the spatial error. Clearly, it is observed that all numerical solutions are in good agreement with exact ones.

To see how good our numerical approach approximates exact solutions, we plot the absolute error, i.e., absolute distance between the exact solution and the numerical approximation for all three models. This shows very good accuracy for our method.

4.1. Effect of Changes in N and S_{max}

To determine the convergence of the discretization scheme, we solve the problem by keeping some parameters fixed and varying others. We fist investigate the effect of the truncation domain on the errors by varying S_{max} between 120 and 500 and keeping other parameters fixed as $K = 100$, $\sigma = 0.2$, $r = 0.03$, $T = 1$, $N = 200$, $S_{min} = 0$, $S_{max} = 250$, $H = 0.6$, $\mu = 0.2$, $k = 1$ and $\alpha = 0.7$. Figure 3 show the L_∞-norm and L_2-norm errors, respectively. We notice that the errors remain bounded and do not vary significantly in term of the truncation domain.

This is an important result, since we can accurately solve the option pricing problem on a small truncated domain, which will result in better efficiency.

In the next experiment, we investigate the convergence of the barycentric Jacobi spectral method. We vary the number of collocation points N between 20 and 300, with parameters $K=100$, $\sigma=0.2$, $r=0.03$, $T=1$, $N=200$, $S_{\min}=0$, $S_{\max}=250$, $H=0.6$, $\mu=0.2$, $k=1$, $\alpha=0.7$; and we plot the results in Figure 4 (top left and right).

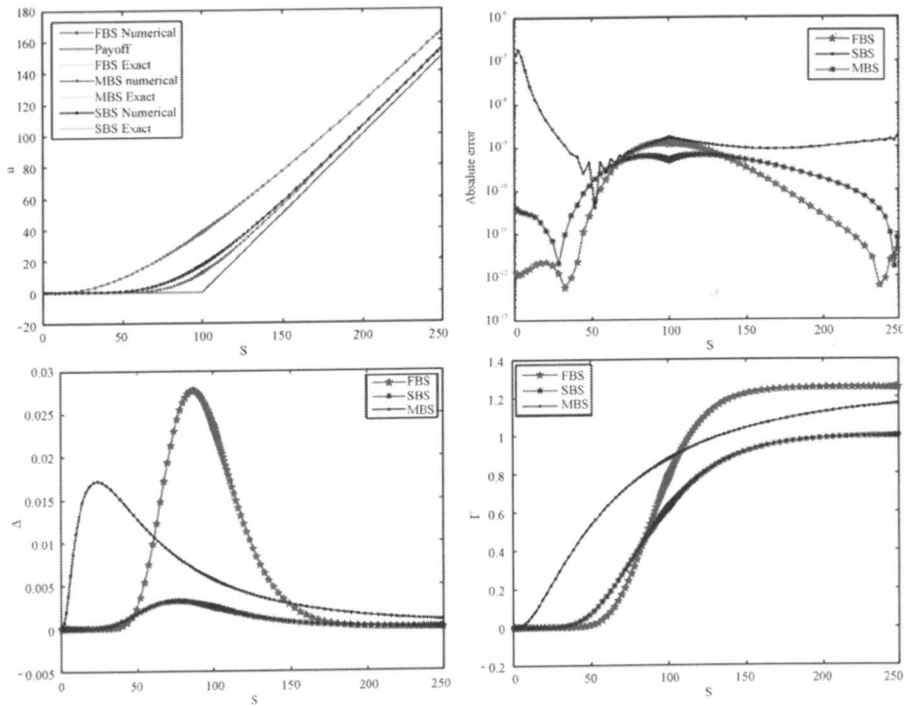

Figure 2. Solutions of the European call options with transaction costs under the three models (top left), absolute error (top right), the Delta (bottom left) and Gamma (bottom right) of the three models with $K=100$, $\sigma=0.2$, $r=0.03$, $T=1$, $N=200$, $S_{\min}=0$, $S_{\max}=250$, $H=0.6$, $\mu=0.2$, $k=1$ and $\alpha=0.7$.

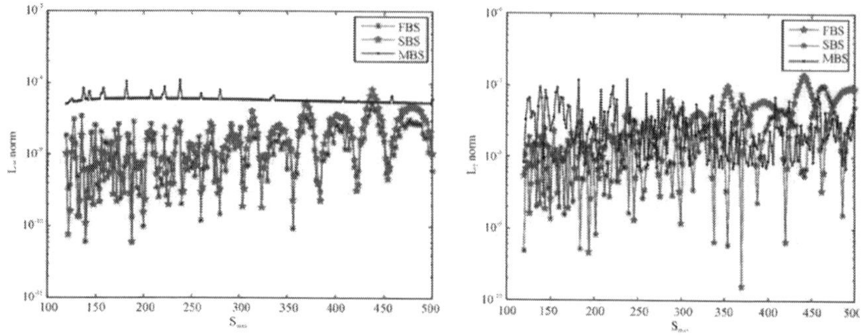

Figure 3. Effect of the truncation domain on the L_2-norm and L_∞-norm in computation of the three models with parameters $K=100$, $\sigma=0.2$, $r=0.03$, $T=1$, $N=200$, $S_{\min}=0$, $S_{\max}=250$, $H=0.6$, $\mu=0.2$, $k=1$ and $\alpha=0.7$.

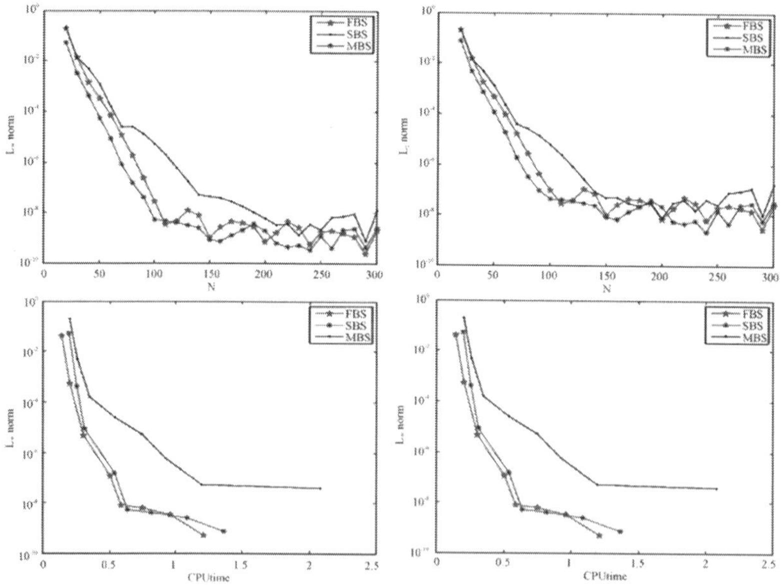

Figure 4. Convergence of L_∞ and L_2-norms and efficiency of the barycentric Jacobi spectral method with parameters $K=100$, $\sigma=0.2$, $r=0.03$, $T=1$, $N=200$, $S_{\min}=0$, $S_{\max}=250$, $H=0.6$, $\mu=0.2$, $k=1$ and $\alpha=0.7$.

For all three models, our method converges rapidly to the exact solutions. This is usually known as spectral or exponential convergence. The efficiency of our approach is measured in Figure 4 (bottom left and right) by the CPU time. The results on the efficiency of our method are very satisfactory for all three methods. An accuracy of 10^{-8} can be obtained in less than a second.

4.2. Convergence of the Method

We explore the effect of changes in time, as well as grid-stretching on the accuracy of the model, keeping

$E=100, \sigma=0.2, r=0.03, T=1, S_{\min}=0, S_{\max}=250, H=0.6, \mu=0.2, k=1, \bar{\alpha}=1, \bar{\beta}=1$

$N=200$ fixed and allowing $\tilde{\alpha}$ to vary between 10^2, 10^3 and 10^4, we compute the L^2- and L_∞-norms for $t=0.5$, $t=1$ and $t=2$. The results are shown in Table1

We notice that when the grid stretching parameter $\tilde{\alpha}$ increases, the accuracy of our method is improved. By accommodating more grid points around the strike price K, we can overcome the poor convergence of naive application of numerical methods when pricing options. We also observe that our method is highly accurate (even for long maturity options).

We also explore the effect of changes in time, as well as Jacobi parameters $\bar{\alpha}$ and $\bar{\beta}$ on the accuracy of the model, keeping $N=200$,
$E=100, \sigma=0.2, r=0.03, T=1, S_{\min}=0, S_{\max}=250, H=0.6, \mu=0.2, k=1, \tilde{\alpha}=$

10^{-4} fixed and allowing $\bar{\alpha}$ and to vary between 0.01, 0.5 and 1, and $\bar{\beta}$ to vary between 0.07, 1 and 2, we compute the L^2- and L_∞-norms for $t=0.5$, $t=1$ and $t=2$. The results are shown in Table2

Once again the barycentric Jacobi spectral method achieves very good accuracy for all three models (even for long term maturity). The Jacobi parameters chosen here do not impact the accuracy of our methodology significantly. It would be useful to investigate how to choose these parameters in an optimal way, however, this is beyond the scope of this paper.

Table 1. Norm infinity and norm relative of errors for the European call options with transaction costs under subdiffusive regime (SBS), fractional Brownian (FBS) and multifractional Brownian (MBS) motions.

Models		L_2			L_∞		
	$\tilde{\alpha}$	$t = 0.5$	$t = 1$	$t = 2$	$t = 0.5$	$t = 1$	$t = 2$
SBS	10^2	6.5320 (−5)	4.6050 (−5)	2.8437 (−5)	6.3396 (−6)	4.2439 (−6)	2.5463 (−6)
	10^3	3.0229 (−8)	2.0966 (−8)	1.2747 (−8)	2.7323 (−9)	1.8275 (−9)	1.0850 (−9)
	10^4	6.6854 (−9)	1.7416 (−8)	3.9353 (−8)	6.8067 (−10)	1.6954 (−9)	3.6965 (−9)
FBS	10^2	9.2345 (−5)	6.5861 (−5)	4.4286 (−6)	9.7058 (−6)	6.5277 (−6)	4.1726 (−6)
	10^3	4.3900 (−8)	3.0709 (−8)	2.0257 (−8)	4.1927 (−9)	2.8170 (−9)	1.7942 (−9)
	10^4	1.0145 (−9)	5.9468 (−9)	1.6573 (−8)	1.6911 (−10)	6.1812 (−10)	1.6266 (−9)
MBS	10^2	2.2002 (−5)	9.8368 (−6)	1.1066 (−5)	1.9491 (−6)	8.6613 (−7)	9.9914 (−5)
	10^3	9.6097 (−9)	1.5761 (−7)	2.7000 (−5)	8.1278 (−10)	1.0433 (−7)	2.6269 (−5)
	10^4	5.2222 (−8)	5.2571 (−7)	5.1523 (−5)	4.7722 (−9)	3.4097 (−7)	5.0791 (−5)

Table 2. Norm infinity and norm relative of errors for the European call options with transaction costs undersubdiffusive regime (SBS), fractional Brownian (FBS) and multifractional Brownian (MBS) motions.

Models			L_2			L_∞		
	$\overline{\alpha}$	$\overline{\beta}$	$t = 0.5$	$t = 1$	$t = 2$	$t = 0.5$	$t = 1$	$t = 2$
	0.01	0.07	1.1292 (8)	1.8273 (−8)	3.3933 (−8)	1.0863 (−9)	1.7419 (−9)	3.1798 (−9)
SBS	0.5	1	1.3716 (−8)	5.1996 (−9)	2.7702 (−8)	1.3623 (−9)	5.2704 (−10)	2.6757 (−9)
	1	2	2.2325 (−8)	1.3673 (−8)	5.6170 (−9)	1.9366 (−9)	1.1587 (−9)	5.5163 (−10)
	0.01	0.07	8.7690 (−9)	1.1102 (−8)	1.8003 (−8)	8.4659 (−10)	1.0787 (−9)	1.7388 (−9)
FBS	0.5	1	2.8309 (−8)	1.4683 (−8)	5.2317 (−9)	2.5996 (−9)	1.4726 (−9)	6.1682 (−10)
	1	2	3.3504 (−8)	2.2705 (−8)	1.4448 (−8)	3.0405 (−9)	2.0069 (−9)	1.2536 (−9)
	0.01	0.07	4.7082(−8)	2.3641 (−7)	1.9080(−5)	4.3052 (−9)	1.3894 (−7)	1.3665 (−5)
MBS	0.5	1	4.2324 (−8)	3.2348 (−7)	5.1813 (−5)	3.8964 (−9)	3.4502 (−7)	5.1095 (−5)
	1	2	7.5002 (−9)	1.4137 (−6)	1.1420 (−4)	5.9867 (−9)	1.3157 (−6)	1.1316 (−4)

5. CONCLUSION

The work of Leland [3] on the discrete option pricing model and replication with transaction cost has paved the way to develop the continuous version. We exploited the continuous version by Wang et al. [8] to construct spectral-based solutions to the model. In practice, option pricing problems are solved numerically since analytical solutions rarely exist. We acknowledge that many studies have been conducted in the domain of finance and both numerical and analytical solutions have been investigated. However, the barycentric Jacobi spectral method has just been proven to have a better accuracy and has not been studied in the field of PDEs in finance. Furthermore, this method obtains solutions with greater accuracy than the usual well-known and studied numerical schemes. The barycentric Jacobi spectral method has full differentiation matrices, which require more computational effort than sparse differentiation matrix methods. In order to improve the efficiency of the rational Jacobi spectral method, we are currently investigating the domain decomposition algorithm which yields block diagonal matrices and we expect to have a greater accuracy, and at least less computational time and memory.

ACKNOWLEDGEMENTS

This work is supported by the Faculty of Natural and Agricultural Science of the University of Pretoria. B. F. Nteumagne and E. Pindza would like to thank Mr. Brad Welch for his financial support in this research.

REFERENCES

1. F. Black and M. Scholes, "Pricing of Options and Corporate Liabilities," Journal of Political Economy, Vol. 81, No. 3, 1973, pp. 637-654.
2. J. C. Hull, "Options, Futures and Other Derivatives," 4th Edition, Prentice-Hall, Upper Saddle River, 2000.
3. H. E. Leland, "Option Pricing and Replication with Transaction Costs," Journal of Finance, Vol. 40, No. 5, 1985, pp. 141-183.
4. P. Boyle and T. Vorst, "Option Replication in Discrete Time with Transaction Costs," Journal of Finance, Vol. 47, No. 1, 1992, pp. 271-293.

5. M. Monoyios, "Option Pricing with Transaction Costs Using a Markov Chain Approximation," Journal of Economic Dynamics and Control, Vol. 28, No. 5, 2004, pp. 889-913.

6. X. T. Wang, "Scaling and Long-Range Dependence in Option Pricing I: Pricing European Option with Transaction Costs under Fractional Black-Scholes Model," Physics A, Vol. 389, No. 3, 2010, pp. 438-444.

7. M. Magdziarz, "Black-Scholes Formula in Subdiffusive Regime," Journal of Statistical Physics, Vol. 136, No. 3, 2009, pp. 553-564.

8. J. Wang, J. R. Liang, L. J. Lv, W. Y. Qui and F, Y. Ren, "Continuous Time Black-Scholes Equation with Transaction Costs in Subdifusive Fractional Brownian Motion Regime," Physics A, Vol. 391, No. 3, 2012, pp. 750-759.

9. H. Gu and Y.-X. Z, "Pricing Option with Transaction Costs under the Subdiffusive Black-Scholes Model" Journal of East China Normal University, Vol. 5, 2012, pp. 85-92.

10. X. T. Wang, E. H. Zhu, M. M. Tang and H. G. Yan, "Scaling and Long-Range Dependence in Option Pricing II: Pricing European Option with Transaction Costs under the Mixed Brownian-Fractional Brownian Model," Physics A, Vol. 389, No. 3, 2010, pp. 445-451

11. R. Baltensperger, J. P. Berrut and B. Noël, "Exponential Convergence of a Linear Rational Interpolant between Transformed Chebyshev Points," Mathematics of Computation, Vol. 68, 1999, pp. 1109-1120.

12. E. Pindza and K. C. Patidar, "Spectral Method for Pricing Options in Illiquid Markets," Proceedings of the International Conference on Numerical Analysis and Applied Mathematics, Vol. 1479, 2012, pp. 1403-1406.

13. E. Ngounda, K. C. Patidar and E. Pindza, "Contour Integral Method for European Options with Jumps," Communications in Nonlinear Science and Numerical Simulation, Vol. 18, No. 3, 2013, pp. 478-492.

14. E. Ngounda, K. C. Patidar and E. Pindza, "A Robust Spectral Method for Solving Heston's Volatility Model" Journal of Optimization Theory and Applications, Springer US, 2013.

15. H. Wang and D. Huybrechs, "Explicit Barycentric Weights for Polynomial Interpolation in the Roots of Extrema of Classical Orthogonal Polynomials," Report TW604, Department of Computer Science, Katholieke Universiteit Leuven, 2012.

16. E. Hairer and G. Wanner, "Solving Ordinary Differential Equations, II: Stiff and Differential-Algebraic Problems," PringerVerlag, New York, 1991.

17. J. L. Lagrange, "Leçons Elémentaires sur les Mathématiques, Données à l'Ecole Normale en 1795," Journal de l'École Polytechnique, Vol. 7, 1795, pp. 183-287.

18. P. Henrici, "Essentials of Numerical Analysis with Pocket Calculator Demonstrations," Wiley, New York, 1982.

19. W. Werner, "Polynomial Interpolation: Lagrange versus Newton," Mathematics of Computation, Vol. 43, 1984, pp. 205-217.

20. J. P. Berrut and L. N. Trefethen, "Barycentric Lagrange Interpolation," SIAM Review, Vol. 46, 2004, pp. 501-517.

21. T. W. Tee and L. N. Trefethen, "A Rational Spectral Collocation Method with Adaptively Transformed Chebyshev Grids Points," Journal of Scientific Computing, Vol. 28, No. 5, 2006, pp. 1798-1811.

9

CHAPTER

On the Efficacy of Fourier Series Approximations for Pricing European Options

A. S. Hurn[1], K. A. Lindsay[1], A. J. McClelland[2]

[1]*School of Economics and Finance, Queensland University of Technology, Brisbane, Australia*

[2]*Numerix Pty Ltd, Sydney, Australia*

ABSTRACT

This paper investigates several competing procedures for computing the prices of vanilla European options, such as puts, calls and binaries, in which the underlying model has a characteristic function that is known in semi-closed form. The algorithms investigated here are the half-range Fourier cosine series, the half-range Fourier sine series and the full-range Fourier series. Their performance is assessed in simulation experiments in which an analytical solution is available and also for a simple affine model of stochastic volatility in which there is no closed-form solution. The results suggest that the half-range sine series approximation is the least effective of the three proposed algorithms. It is rather more difficult to distinguish between the performance of the half- range cosine series and the full-range Fourier series. However there are two clear differences. First, when the interval over which the density is approximated is relatively large, the full-range Fourier series is at least as good as the half-range Fourier cosine series, and outperforms the latter in pricing out-of-the-money call options, in particular with maturities of three months or less. Second, the computational time required by the half-range Fourier cosine series is uniformly longer than that required by the full-range Fourier series for an interval of fixed length. Taken together, these two conclusions make a case for pricing options using a full-range range Fourier series as opposed to a half-range Fourier cosine series if a large number of options are to be priced in as short a time as possible.

Keywords: Fourier Transform, Fourier Series, Characteristic Function, Option Price

1. INTRODUCTION

The need to price vanilla European options in a rapid manner arises in numerous activities at financial institutions. Perhaps the most well-known situation in which this need occurs in practice is model calibration, in which exotic options are priced using models with values for the (risk-neutral) parameters chosen in such a way as to ensure that the model reproduces quoted prices for liquid options. For each set of parameters considered in the search space, it is therefore necessary to evaluate the prices of all options in the calibration set using the model, before comparing these with the quoted prices. In a similar vein, there is a growing literature which uses large panels of options data for estimating model parameters ([1] -[5]) in which the computational complexity is driven by the evaluation of vast numbers of option prices. Consequently, this paper investigates several competing procedures for computing the prices of vanilla European options, such as puts, calls and binaries and assesses their comparative performance on the basis of accuracy and computational speed.

Various strategies have been proposed for calculating the price of option contracts from knowledge of the conditional characteristic function of the underlying model. It is an important fact that a surprisingly large number of models have a semi-closed expression for their conditional characteristic function. For example, the identification of the conditional characteristic function for multivariate affine models with/without jump processes leads to the solution of a family of ordinary differential equations, albeit in the complex plane. In view of the Levy-Khintchine theorem, the identification of the conditional characteristic function for Levy processes is expressed in terms of various integrals with respect to the Levy measure.

The most commonly used techniques for taking advantage of a known conditional characteristic function have at their core the application of the Fast Fourier Transform (FFT). The best documented approaches is due to Carr and Madan [6] who construct an expression for the price of a European call option in terms of an integral over the characteristic function. This integral, which has an oscillatory kernel, is computed by an application of the FFT. Borak, Detlefsen and Hardle [7] apply the FFT strategy and demonstrate its efficacy by comparison with Monte Carlo simulation for a variety of models. Lord, Fang, Bervoets and Oosterlee [8] and Kwok, Leung and Wong [9] demonstrate how Fourier's convolution theorem in combination with the FFT can be used to price certain exotic options from knowledge of the conditional characteristic function of the price of the underlying asset. A different approach pioneered by Fang and Oosterlee [10] uses the characteristic function to directly approximate the marginal transitional probability density of returns by a Fourier cosine series. Recently Zhang, Grzelak and Oosterlee [11] have demonstrated how this methodology can be extended to the pricing of early-exercise commodity options under the Ornstein-Uhlenbeck process.

Rather than describing in detail the nuances of these various strategies, it is useful to point out what overarching assumptions connect them. Recall that the FFT is simply a clever piece of linear algebra that reduces the arithmetical load in implementing the Discrete Fourier Transform (DFT), namely the pair of equations connecting the coefficients of a finite Fourier series with values of the underlying function and vice versa. Therefore the decision to use the FFT implicitly makes the assumption that the underlying function is periodic over an interval of finite length, in practice determined by the range of frequencies submitted to the characteristic function, and that the function has been approximated over the interval by a finite Fourier series. The values of Fourier coefficients calculated from the characteristic function are in error by the extent to which the Finite Fourier Transform[1] differs from the Fourier transform.

Thus techniques using the FFT and those based on the construction of Fourier series share the same common assumptions and deficiencies. However, an important difference between an implementation using the FFT approach and one using the Fourier series approach is that the latter is parsimonious in its use of arithmetic whereas the former typically performs more arithmetic than necessary, albeit in an efficient way. For example, if the FFT is used to determine the value of a probability density function, what is recovered is the value of the function at each node of the interval, whereas all that might be needed is the value of the probability density function over a sub-interval.

The focus of this work is on the algorithm proposed by Fang and Oosterlee [10] who give a convincing demonstration of the efficacy of the Fourier cosine series. This series is more accurately called the half-range[2] cosine series because the actual function to be expanded is defined only on half the interval of periodicity (or range), the function being extended to the full range as an even-valued function. Half-range cosine series usually fail to represent derivatives whereas half-range Fourier sine series usually fail to represent function values. While the use of the half-range Fourier cosine series is a solid idea, Fang and Oosterlee [10] provide no motivation or explanation as to why this choice of approximating transitional density should be preferred over the half-range Fourier sine series or the full-range Fourier series for that matter. For example, intuition would suggest that the latter might perform better simply because it uses higher frequencies, which in turn translate to a more rapidly converging Fourier expansion. Indeed this intuition is borne out in calculation, but of course speed is not the only criterion of relevance in assessing the efficacy of a numerical procedure.

An important but subtle difference shared by the half-range Fourier cosine and full-range Fourier series approximations of density, but different from representations of density based on the half-range Fourier sine series, is that the former assign unit probability to the interval of support when in reality probability lies outside this interval, whereas the latter imposes zero probability density at the endpoints of the interval of support in contravention of reality, but on the other hand does not assign unit probability to the interval of support. Is one approach always superior to the other or is it a case of horses for courses? Intuition might suggest the latter. For example, when pricing a call option the most important component of the pricing error comes from the exclusion of contributions from asset price exterior to the finite interval of support. Because the half-range cosine and full-range Fourier series necessarily capture unit density, intuition might suggest that these approximations provide potential compensation for this component of pricing error. On the other hand intuition would suggest that the same approximations, when used to price binary options, might have a tendency to exaggerate the probability of exercise and therefore overprice this option in contrast to the half-range sine series approximation of probability density.

2. FOURIER SERIES AND TRANSFORM

Suppose that $f(y)$ satisfies the Dirichlet conditions on $[a,b]$, then there are three common ways in which $f(y)$ may be represented by a Fourier series. These are the half-range Fourier cosine series, the half-range Fourier sine series and the full-range Fourier series with respective representations

$$\text{(a)}\quad \frac{a_0}{2}+\sum_{k=1}^{\infty} a_k \cos\left(\frac{k\pi(y-a)}{b-a}\right),$$
$$\text{(b)}\quad \sum_{k=1}^{\infty} b_k \sin\left(\frac{k\pi(y-a)}{b-a}\right),$$
$$\text{(c)}\quad \frac{a_0}{2}+\sum_{k=1}^{\infty} a_k \cos\left(\frac{2k\pi(y-a)}{b-a}\right)+b_k \sin\left(\frac{2k\pi(y-a)}{b-a}\right). \qquad (1)$$

The use of the term "half-range" in describing Expressions (a) and (b) simply refers to the fact that the function $f(y)$, although defined in $[a,b]$, has for the construction of the Fourier series been extended into the interval $[2a-b,a]$ as an even-valued function in the case of the half-range cosine series (so that sine contributions vanish) and as an odd-valued function in the case of the half-range sine series (so that the constant and cosine contributions vanish). Thus both half-range series are conventional Fourier series taken over the interval $[2a-b,b]$ such that the function represented by the half-range Fourier cosine series is usually not differentiable at $y=a$, whereas that represented by the half-range Fourier sine series is usually discontinuous at $y=a$.

In the case of the half-range cosine and sine series in Expressions (1a) and (1b) respectively, the coefficients $a_k\ (k\geq 0)$ and $b_k\ (k\geq 1)$ are calculated from the function $f(y)$ via the formulae

$$a_k=\phi\int_a^b f(y)\cos\left(\frac{k\pi(y-a)}{b-a}\right)\mathrm{d}y$$
$$b_k=\phi\int_a^b f(y)\sin\left(\frac{k\pi(y-a)}{b-a}\right)\mathrm{d}y, \qquad (2)$$

in which

$$\phi = \frac{2}{b-a}.$$

In terms of the exponential function the real coefficients a_k and b_k are calculated respectively as the real and imaginary parts of the single complex Equation

$$a_k + ib_k = \phi \int_a^b f(y) \exp\left(\frac{k\pi i (y-a)}{b-a} \right). \tag{3}$$

In the case of the full-range Fourier series in Expression (1c) the real coefficients $a_k \ (k \geq 0)$ and $b_k \ (k \geq 1)$ are calculated respectively as the real and imaginary parts of the complex Equation

$$a_k + ib_k = \phi \int_a^b f(y) \exp\left(\frac{2k\pi i (y-a)}{b-a} \right) \mathrm{d}y. \tag{4}$$

Suppose now that $f(y)$ is a transitional probability density function with known characteristic function defined formally by the Equation

$$\chi(\omega) = \mathbb{E}\left[\mathrm{e}^{i\omega y} \right] = \int_{\mathbb{R}} f(y) \mathrm{e}^{i\omega y} \, \mathrm{d}y, \tag{5}$$

where $\omega \in \mathbb{R}$ and $\chi(0) = 1$. A necessary condition for $f(y)$ to be a probability density function is that $f(y) \to 0$ as $|y| \to \infty$, and therefore there is guaranteed to be an interval $[a,b]$ such that for all $y \in (-\infty, a] \cup [b, \infty)$ it can be asserted that $f(y) < \varepsilon$ for any arbitrary small positive ε. The implication of this observation is that the Fourier coefficients in Equations (3) and (4) can be approximated from knowledge of the characteristic function via the respective formulae $a_k \approx A_k$ and $b_k \approx B_k$, where

$$A_k = \phi\Re\left[\chi\left(\frac{k\pi}{b-a}\right)\exp\left(\frac{-k\pi ia}{b-a}\right)\right],$$
$$B_k = \phi\Im\left[\chi\left(\frac{k\pi}{b-a}\right)\exp\left(\frac{-k\pi ia}{b-a}\right)\right], \tag{6}$$

while the coefficients of the full-range Fourier series can be approximated from knowledge of the characteristic function via the formula $a_k \approx A_k$ and $b_k \approx B_k$, where

$$A_k + iB_k = \phi\chi\left(\frac{2k\pi}{b-a}\right)\exp\left(\frac{-2k\pi ia}{b-a}\right). \tag{7}$$

The accuracy of approximations (6) and (7) is investigated in Section 6, where it is demonstrated that the error can be made arbitrarily small by choosing a suitably large interval.

3. APPROXIMATING PROBABILITY DENSITY FUNCTIONS

The quality of this practical idea is now explored for three trial probability density functions with known closed-form expressions for their characteristic functions. The first choice is the Gaussian density which may be regarded as representative of distributions with super-exponentially decaying tail density. The second and third choices are the Gamma density and the Cauchy density which are treated as representative examples of distributions with exponentially decaying and algebraically decaying tail densities respectively.

3.1. Gaussian Density

It is well known that the Gaussian density with mean value μ and variance σ^2 has characteristic function $\chi(\omega)=\exp(i\mu\omega-\sigma^2\omega^2/2)$. In order to demonstrate the quality with which the true probability density $f(x)$ can be reconstructed from a truncated Fourier series of the form of Equation (1c), suppose that $\mu=\sigma=1$ and take the interval of support to be $[a,b]=[-3,5]$, i.e. four standard deviations on either side of the mean value. The approximating function in this case using $N/2$ frequencies is

$$\hat{f}(x)=\frac{1}{8}\left[c_0+2\sum_{n=1}^{N/2}(c_n\cos k_n x+s_n\sin k_n x)\right], \tag{8}$$

where $k_n=n\pi/4$. Figure 1 illustrates the quality of this approximation using 40 frequencies (solid line, $N=80$) and using 4 frequencies (dashed line, $N=8$).

With as few as 4 frequencies it is clear that the approximating density still provides a good representation of the true density; with 40 frequencies the approximating density function is indistinguishable from the true density function, at least in terms of the resolution in Figure 1. It will be seen that this excellent performance is explained by the fact that the cumulative distribution function of the Gaussian probability density function converges to zero super-exponentially as $x\to-\infty$ and to unity as $x\to\infty$.

3.2. Gamma Density

The Gamma density with shape parameter α and scale parameter β has probability density function and characteristic function give by the respective formulae

$$f(x)=\frac{1}{\Gamma(\alpha)\beta}\left(\frac{x}{\beta}\right)^{\alpha-1} e^{-x/\beta},$$
$$\chi(\omega)=\frac{1}{(1-i\beta\omega)^{\alpha}}. \tag{9}$$

The approximating density is identical to Expression (8) with $b-a=8$. The task is now to reconstruct the Gamma probability density function from $\chi(\omega)$. Because $Y=X/\beta$ is also Gamma distributed, in this case with shape parameter α and unit scale parameter, then the appropriate comparison for the Gamma distribution is between the approximating Fourier series and a Gamma distribution with unit scale parameter. Figure 2 illu-

strates the quality of this approximation for $\Gamma\left(\frac{3}{2},1\right)$ for 50 frequencies (solid line, $N=100$) and 20 frequencies (dashed line, $N=40$). With the exception of the region very close to the origin, the true density and approximated density are not significantly different for even 20 frequencies, and with 50 frequencies the difference between the true density and the approximating density is not discernible with the exception of the origin which does not present a difficulty since the density to known to be zero there.

The quality of the approximation is again due to the fact that the cumulative distribution function of the Gamma density converges exponentially to zero as $x \to \infty$. Typical values for the coefficient of mean reversion (say, $\kappa=3.0$), the mean volatility (say $\gamma=0.02$) and the volatility of volatility (say $\sigma=0.2$) in Heston's model of stochastic volatility lead to a stationary distribution of volatility described by a Gamma density with shape parameter $\alpha = 2\kappa\gamma/\sigma^2 = 3.0$. A second example with $\alpha = 3$ and $\beta = 1$ is illustrated in Figure 3.

The important observation from both of these experiments is that distributions with exponentially decaying tail density can also be well described by a relatively small range of frequencies.

3.3. Cauchy Density

The Cauchy density with median μ and scale parameter α has probability density function and characteristic function give by the respective formulae

$$f(x) = \frac{\alpha}{\pi} \frac{1}{(x-\mu)^2 + \alpha^2},$$
$$\chi(\omega) = e^{i\mu\omega - \alpha|\omega|}. \tag{10}$$

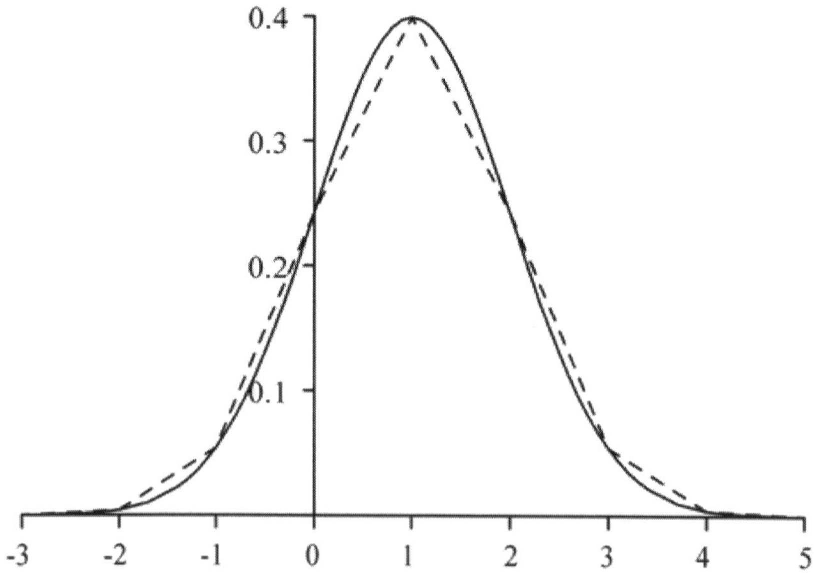

Figure 1. Comparison of the true Gaussian density and its approximation based on 40 frequencies (solid line, N = 80) and 4 frequencies (dashed line, N = 8).

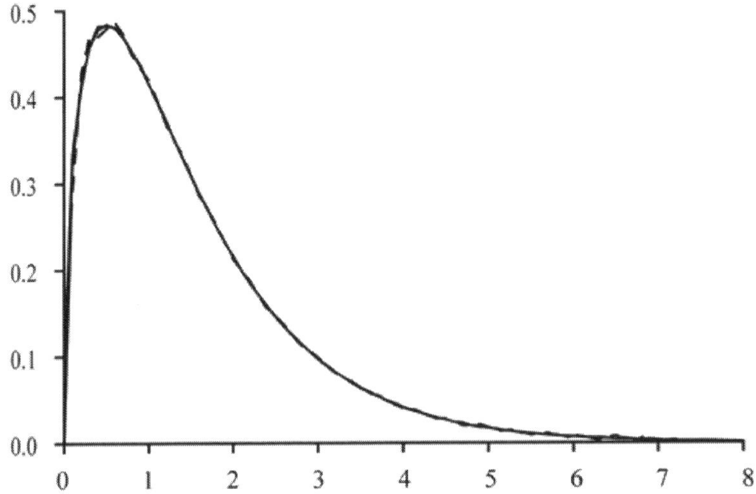

Figure 2. Comparison of the Gamma density $\Gamma(3/2,1)$ and its approximations based on 50 frequencies (solid line, N = 100) and 20 frequencies (dashed line, N = 40).

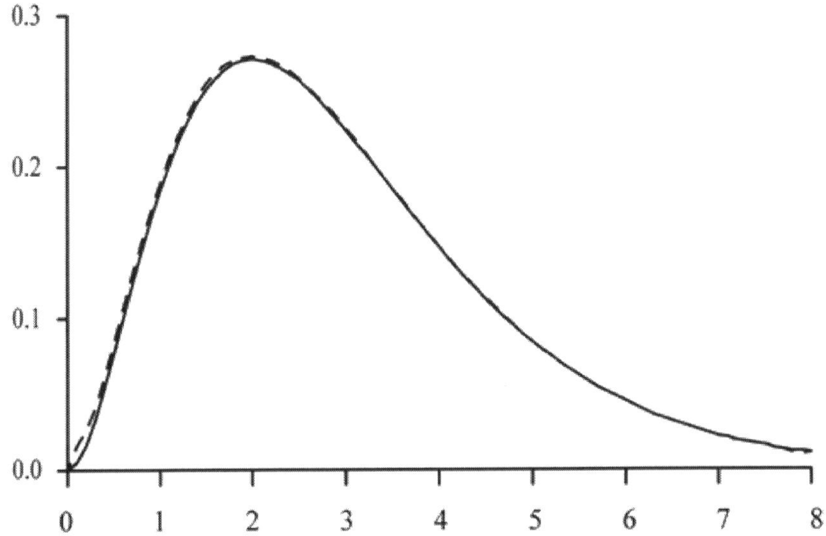

Figure 3. Comparison of the Gamma density $\Gamma(3,1)$ (solid line) and its approximations based on 10 frequencies (dashed line, N = 20).

The approximating density is again Expression (8) with $b-a=20$. The task is now to reconstruct the Cauchy probability density function from $\chi(\omega)$, and for this purpose let $\mu=2$, $\alpha=1$ and suppose that $f(x)$ is treated as a function of compact support over the interval $[a,b]=[-8,12]$. The cumulative dis- tribution function of the Cauchy density is

$$F(x)=\frac{1}{2}+\frac{1}{\pi}\tan^{-1}\left(\frac{x-\mu}{\alpha}\right),$$

which in this case converges algebraically to zero as $x\to-\infty$ and algebraically to unity as $x\to\infty$. Figure 4 illustrates the impact of this inferior rate of decay for the approximation based on 10 frequencies.

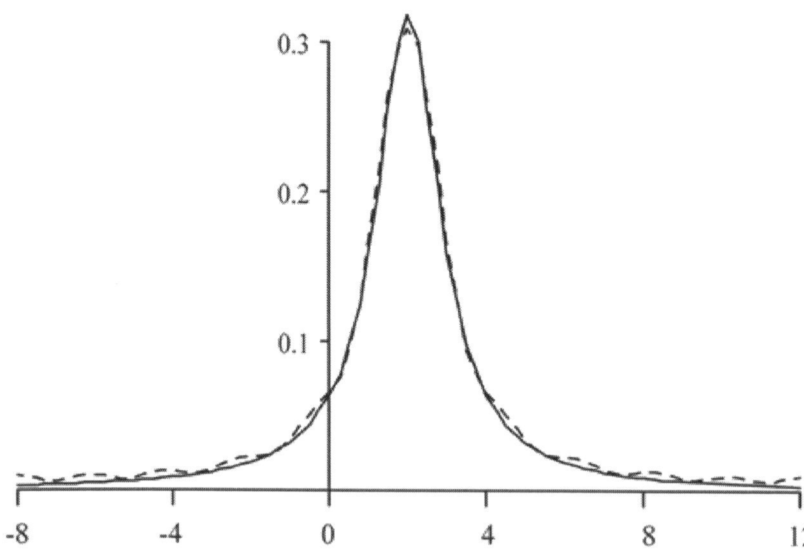

Figure 4. Comparison of the Cauchy density with parameters $\mu=2$ and $\alpha=1$ (solid line) and the approximating Fourier series based on 10 frequencies (dashed line, N = 20).

While some erratic behaviour is evident in the tails of the approximating density, nevertheless the quality of the approximation is remarkably good considering the small number of frequencies in use.

In general, the approximate representations of the Gaussian, Gamma and Cauchy probability density functions all share the common fact that $F(a)+1-F(b)$ provides an upper bound for the error in the representation of density, where $F(x)$ is the conventional cumulative distribution function of the underlying probability density function $f(x)$. In reality the error is significantly less than this upper bound due to cancellations resulting from the fact that the integrand in each of the Gaussian, Gamma and Cauchy densities is the product of a slowly varying probability density function decaying to zero and an independent rapidly oscillating function (in this case trigonometric functions represented in exponential form). However, the presence of this cancellation cannot change the character of the error which is entirely determined by the properties of $F(x)$ as $x \to -\infty$ and as $x \to \infty$.

3.4. Comparison of True and Approximating Densities

While the illustrations of Figures 1-4 suggest that the Fourier series approximation of density is effective, it would be useful to quantify just how well density is approximated by the various Fourier methods. Instinctively it would seem that one possible way to achieve this objective is to use the Kullback-Leibler (KL) divergence criterion

$$D = \int_{\mathbb{R}} p(x) \log \frac{p(x)}{q(x)} dx \tag{11}$$

to measure the "distance" or departure of the probability density function $q(x)$ from that of $p(x)$, which in the formulation of Equation (11) is taken to be the true probability density function. The measure appeals to the fact that $D \geq 0$ with equality if and only if the density functions $p(x)$ and $q(x)$ are identical. However, the KL criterion is simply an application of Jensen's inequality for the concave function $\log x$, and the properties of the criterion rely critically on the fact that $p(x)$ and $q(x)$ are probability densities, i.e. they are non-negative functions which integrate to unity.

The central idea of each Fourier approximation is to replace the true probability density function by a function of compact support, that is, $q(x) \equiv 0$ outside its interval of support. Consequently using measure (11) to compute the departure of the approximating density $q(x)$ from the true density $p(x)$ will typically require division by zero thereby rendering inappropriate the application of the KL criterion when used with $p(x)$ as the true probability density. However, inverting the problem, one can use the KL criterion to measure the distance between the approximating density $q(x)$ and the true density $p(x)$, that is, one may compute the value of D in Expression (11) taking $p(x)$ to be the approximating density and $q(x)$ to be the true density. In this case the integrand of expression (11) is well defined for all values of x, and therefore in order to proceed in this direction it remains to ensure that the approximating density integrates to unity. Approximations of density based on either the half range Fourier cosine series or the full range Fourier series are guaranteed to have this property, whereas an approximation based on the half range Fourier sine series need not share this property and therefore is inappropriate for analysis using the KL criterion. Tables 1-3 show values for the KL criterion for the Gaussian distribution with mean value one and unit standard deviation, the Gamma distribution with shape parameter 3 and scale parameter one, and the Cauchy distribution with median two and scale parameter one.

As might be anticipated, the Gaussian distribution is most efficiently approximated by Fourier methods followed by the Gamma distribution and finally the Cauchy distribution. However, it is clear that all of these distributions are well approximated by the half range Fourier cosine and full range Fourier series for sufficiently large intervals of support and an adequate number of frequencies. The results in these tables also reinforce the idea that the number of frequencies used in the approximation is of secondary importance to the size of the interval of support once sufficient frequencies are in use. This observation accords with intuition in the respect that larger intervals of support capture more of the true density and reduce the distortion in the approximating density, which as has already been commented, will always integrate to one. It would seem that approximations based on 50 frequencies are adequate in all of these examples. The results suggest that using more frequencies provides no meaningful improvement in accuracy. Moreover, for practical purposes there is little to choose between approximation based on the half range Fourier cosine and the full range Fourier series, although the latter has a slight edge for sufficiently large intervals and an adequate number of frequencies.

Table 1. Values of the KL measure (D_c for the half range Fourier cosine series and D_f for the full Fourier series) are given for the Gaussian density with unit mean and unit standard deviation. Values measure the deviation of the true density deviates from the approximating density for intervals of length 6, 8, 10, 12 and 14 standard deviations centred about the mean of the Gaussian distribution using 5, 10, 25, 50, 100 and 200 frequencies.

Interval	KL	Number of frequencies used in Fourier sums					
Length	Measure	5	10	25	50	100	200
6	D_c	0.0038	0.0031	0.0031	0.0031	0.0031	0.0031
	D_f	0.0031	0.0031	0.0031	0.0031	0.0031	0.0031
8	D_c	0.0498	0.0001	0.0001	0.0001	0.0001	0.0001
	D_f	0.0001	0.0001	0.0001	0.0001	0.0001	0.0001
10	D_c	0.4284	0.0005	0.0000	0.0000	0.0000	0.0000
	D_f	0.0005	0.0000	0.0000	0.0000	0.0000	0.0000
12	D_c	1.2491	0.0181	0.0000	0.0000	0.0000	0.0000
	D_f	0.0181	0.0000	0.0000	0.0000	0.0000	0.0000
14	D_c	2.4615	0.1301	0.0000	0.0000	0.0000	0.0000
	D_f	0.1301	0.0000	0.0000	0.0000	0.0000	0.0000

Table 2. Values of the KL measure (D_c for the half range Fourier cosine series and D_f for the full Fourier series) are given for the Gamma density with shape parameter 3 and unit scale factor. Values measure the deviation of the true density deviates from the approximating density for intervals [0, L] of length 6, 8, 10, 12 and 14 scale factors using 5, 10, 25, 50, 100 and 200 frequencies.

Interval	KL	Number of frequencies used in Fourier sums					
Length	Measure	5	10	25	50	100	200
6	D_c	0.0793	0.0720	0.0715	0.0715	0.0715	0.0715
	D_f	0.0890	0.0874	0.0871	0.0871	0.0871	0.0871
8	D_c	0.0396	0.0174	0.0158	0.0158	0.0158	0.0158
	D_f	0.0196	0.0173	0.0168	0.0168	0.0167	0.0167
10	D_c	0.0669	0.0076	0.0033	0.0032	0.0032	0.0032
	D_f	0.0105	0.0039	0.0031	0.0031	0.0031	0.0031
12	D_c	0.1532	0.0142	0.0008	0.0006	0.0006	0.0006
	D_f	0.0290	0.0038	0.0007	0.0006	0.0005	0.0005
14	D_c	0.2944	0.0391	0.0006	0.0001	0.0001	0.0001
	D_f	0.0745	0.0111	0.0006	0.0001	0.0001	0.0001

Table 3. Values of the KL measure (D_c for the half range Fourier cosine series and D_f for the full Fourier series) are given for the Cauchy density with median two and unit scale factor. Values measure the deviation of the true density deviates from the approximating density for intervals of length 14, 20, 26, 32 and 38 scale factors centred about the median of the Cauchy distribution using 5, 10, 25, 50, 100 and 200 frequencies.

Interval	KL	Number of frequencies used in Fourier sums					
Length	Measure	5	10	25	50	100	200
14	D_c	0.2425	0.1148	0.1076	0.1075	0.1075	0.1075
	D_f	0.1148	0.1076	0.1075	0.1075	0.1075	0.1075
20	D_c	0.4117	0.1296	0.0760	0.0753	0.0753	0.0753
	D_f	0.1296	0.0775	0.0753	0.0753	0.0753	0.0753
26	D_c	0.6356	0.2188	0.0630	0.0580	0.0580	0.0580
	D_f	0.2188	0.0713	0.0580	0.0580	0.0580	0.0580
32	D_c	0.8602	0.3391	0.0670	0.0472	0.0471	0.0471
	D_f	0.3391	0.0932	0.0472	0.0471	0.0471	0.0471
38	D_c	1.0715	0.4885	0.0937	0.0403	0.0397	0.0397
	D_f	0.4885	0.1423	0.0403	0.0397	0.0397	0.0397

4. PRICING EUROPEAN OPTIONS

The successfully pricing of European option contracts for affine models of stochastic volatility requires knowledge of the marginal density of the asset price under the risk-neutral measure. The difficulty, however, is that no closed-form expression for this density is available for even the simplest of the multivariate affine models used in finance, although it is well known that such models have characteristic function, $\chi(t,\omega)$, of generic form

$$\exp\left[\beta_0(\tau,\omega)+\beta_1(\tau,\omega)Y+\sum_{k=2}^{M}\beta_k(\tau,\omega)X_k\right], \tag{12}$$

where T is the maturity of the option, $\tau=T-t$ is the backward variable and ω is the characteristic variable associated with the non-dimensional variable Y, here defined to be the logarithm of the ratio of asset price to strike price. The state variables $X=(X_2,\cdots,X_M)$ in Expression (12) denote the values of the latent states of the system at time t. Typically the functions $\beta_0,\cdots,\beta_M$ are the solution of a system of $(M+1)$ ordinary differential Equations with initial conditions $\beta_1(0,\omega)=i\omega$ and $\beta_0(0,\omega)=\beta_2(0,\omega)=\cdots=\beta_M(0,\omega)=0$. In overview, there is a well trodden procedure that starts with the specification of the multivariate affine model and ends with the construction of $\chi(t,\omega)$.

4.1. European Call Option

In the case of a European call option with strike price K and maturity T on an asset with spot price S_0, the price of the option is

$$e^{-rT}\int_K^{\infty}(S-K)\tilde{f}_Q(S_0,X_2\cdots X_M,T|S;\theta)dS,$$

where $\tilde{f}_{\mathrm{Q}}\left(S_0, X_2 \cdots X_M, T \mid S; \theta\right)$ is the marginal density of the asset price at the maturity of the option when S_0 is the spot price of the asset and $\left(X_2, \cdots, X_M\right)$ are the spot values of the latent states. When expressed in terms of $y = \log\left(S/K\right)$, the price of the option, say C, becomes

$$\mathrm{e}^{-rT} K \int_0^{\infty} \left(\mathrm{e}^y - 1\right) f_{\mathrm{Q}}\left(\xi, X_2 \cdots X_M, T \mid y; \theta\right) \mathrm{d}y, \tag{13}$$

where $\xi = \log\left(S_0/K\right)$. The value of this integral is now approximated under the assumption that $f_{\mathrm{Q}}\left(\xi, X_2 \cdots X_M, T \mid y; \theta\right)$ is well approximated by a function of compact support over the interval $[a, b]$. Of course, the specific form taken by this approximation will depend of the choice of Expressions (a), (b) or (c) in Equations (1), but in each case the approximation of $f_{\mathrm{Q}}\left(\xi, X_2 \cdots X_M, T \mid y; \theta\right)$ by a function of compact support in $[a, b]$ necessarily changes the interval of integration in Expression (13) from $[0, \infty)$ to $[0, b]$. Thereafter it follows that the costs, C, of a call option on the basis of approximations (a) and (b) are respectively

$$\begin{aligned}
&(\mathrm{a}) \quad C = \mathrm{e}^{-rT} K \left[\frac{a_0}{2}\left(\mathrm{e}^b - b - 1\right) + \sum_{k=1}^{\infty} a_k \frac{\omega_k \mathrm{e}^b (-1)^k - \omega_k \cos \omega_k a - \sin \omega_k a}{\omega_k \left(1 + \omega_k^2\right)} \right], \\
&(\mathrm{b}) \quad C = \mathrm{e}^{-rT} K \sum_{k=1}^{\infty} \frac{b_k}{\omega_k} \left[\left(\mathrm{e}^b - 1\right)(-1)^{k-1} + \frac{\omega_k \sin \omega_k a - \cos \omega_k a + \mathrm{e}^b (-1)^k}{1 + \omega_k^2} \right],
\end{aligned} \tag{14}$$

where $\omega_k = k\pi/(b - a)$. Based on approximation (c) the price of a call option is

$$\text{(c)}\quad C = e^{-rT}K\left[\frac{a_0}{2}\left(e^b - b - 1\right) + \sum_{k=1}^{\infty} a_k \frac{\lambda_k e^b - \lambda_k \cos\lambda_k a - \sin\lambda_k a}{\lambda_k\left(1+\lambda_k^2\right)}\right.$$
$$\left. + \frac{b_k}{\lambda_k}\left(1 - e^b + \frac{\lambda_k \sin\lambda_k a - \cos\lambda_k a + e^b}{1+\lambda_k^2}\right)\right], \qquad (15)$$

where $\lambda_k = 2k\pi/(b-a)$. The primary computational load in the computation of Expressions (a), (b) and (c) resides in the calculation of trigonometric functions, and consequently individual terms of Expression (c) require more arithmetical effort than the equivalent terms of either Expression (a) or (b). However the difference in computational load is minor when the more rapid convergence of summation (c) is compared with that of summations (a) and (b) simply because the frequencies used in approximation (c) are exactly double those used in approximations (a) and (b).

4.2. Binary Option

The price of a binary (call) option with strike price K and maturity T on an asset with spot price S_0 is

$$e^{-rT}\int_K^{\infty} \tilde{f}_Q\left(S_0, X_2 \cdots X_M, T \mid S; \theta\right) dS,$$

where $\tilde{f}_Q\left(S_0, X_2 \cdots X_M, T \mid S; \theta\right)$ is the marginal density of the asset price at the maturity of the option when the asset has spot price S_0 and $(X_2, \cdots, X_M)$ are the spot values of the latent states. When expressed in terms of $y = \log(S/K)$, the price of the binary becomes

$$D = e^{-rT}\int_0^{\infty} f_Q\left(\xi, X_2 \cdots X_M, T \mid y; \theta\right) dy, \qquad (16)$$

where $\xi = \log(S_0/K)$. The value of this integral is now approximated under the assumption that $f_Q(\xi, X_2 \cdots X_M, T | y;\theta)$ is well represented by the procedures proposed in Section 2. Specifically, the price of a binary option computed from the half-range Fourier cosine and sine series are respectively

$$\text{(a)} \quad D = e^{-rT}\left[\frac{a_0}{2}b + \sum_{k=1}^{\infty} a_k \frac{\sin \omega_k a}{\omega_k}\right],$$

$$\text{(b)} \quad D = e^{-rT}\sum_{k=1}^{\infty} \frac{b_k\left(\cos \omega_k a - (-1)^k\right)}{\omega_k}, \tag{17}$$

where $\omega_k = k\pi/(b-a)$. The price of the binary option based on the full-range Fourier series (c) is

$$\text{(c)} \quad D = e^{-rT}\left[a_0 b + \sum_{k=1}^{\infty} a_k \frac{\sin \lambda_k a}{\lambda_k} + \sum_{k=1}^{\infty} \frac{b_k}{\lambda_k}(\cos \lambda_k a - 1)\right], \tag{18}$$

where $\lambda_k = 2k\pi/(b-a)$. As is the case with the pricing of a call option, the frequencies used in approximation (c) are double those used in approximations (a) and (b) suggesting that calculations based on Expression (c) can be expected to converge more rapidly than those based on Expressions (a) and (b).

5. THE HESTON MODEL OF STOCHASTIC VOLATILITY

Heston's [12] risk-neutral model of stochastic volatility has expression

$$dY = (r - q - V/2)dt + \sqrt{V}\left(\sqrt{1-\rho^2}dW_1 + \rho dW_2\right),$$
$$dV = \kappa_Q(\gamma_Q - V)dt + \sigma\sqrt{V}dW_2, \tag{19}$$

where $Y = \log S/K$, V is the diffusion of asset price, r is the risk-free rate of interest, q is the dividend rate, κ_Q is the risk-neutral rate of mean reversion of volatility to the risk-neutral long run value of γ_Q, σ scales the volatility of diffusion, ρ is the local correlation between returns and volatility and dW_1, dW_2 are increments in the independent Brownian motions W_1 and W_2.

Suppose that $f(Y,V,t|y,v,T)$ is the probability density function of Equations (19) expressed in terms of the backward state (Y,V) and the forward state (y,v), then the backward Kolmogorov Equation satisfied by the transitional probability density function of Equations (19) is

$$\frac{\partial f}{\partial t} + (r - q - V/2)\frac{\partial f}{\partial Y} + \kappa_Q(\gamma_Q - V)\frac{\partial f}{\partial V} + \frac{V}{2}\left(\frac{\partial^2 f}{\partial Y^2} + 2\rho\sigma\frac{\partial^2 f}{\partial Y \partial V} + \sigma^2\frac{\partial^2 f}{\partial V^2}\right) = 0. \tag{20}$$

Let $\chi(Y,V,t,\omega_Y,\omega_V)$ be the characteristic function of $f(Y,V,t|y,v,T)$ with respect to the forward variables, that is,

$$\chi(Y,V,t,\omega_Y,\omega_V) = \int_{Q^2} f(Y,V,t|y,v,T)e^{i(\omega_Y y + \omega_V v)}dydv. \tag{21}$$

By taking the Fourier transform of Equation (20) with respect to the backward variables, the characteristic function $\chi(Y,V,t,\omega_Y,\omega_V)$ is seen to satisfy the partial differential Equation

$$\frac{\partial\chi}{\partial t}+(r-q-V/2)\frac{\partial\chi}{\partial Y}+\kappa_Q(\gamma_Q-V)\frac{\partial\chi}{\partial V}+\frac{V}{2}\left(\frac{\partial^2\chi}{\partial Y^2}+2\rho\sigma\frac{\partial^2\chi}{\partial Y\partial V}+\sigma^2\frac{\partial^2\chi}{\partial V^2}\right)=0. \tag{22}$$

with terminal condition

$$\chi(Y,V,T,\omega_Y,\omega_V)=\exp\left[i(\omega_Y Y+\omega_V V)\right].$$

Thereafter, it is straightforward to show that the ansatz

$$\chi(Y,V,t,\omega_Y,\omega_V)=e^{\beta_0(\tau)+\beta_1(\tau)Y+\beta_2(\tau)V} \tag{23}$$

with $\tau=T-t$ is a solution of Equation (22) provided the coefficient functions $\beta_0(\tau)$, $\beta_1(\tau)$ and $\beta_2(\tau)$ satisfy the ordinary differential equations

$$\begin{aligned}\frac{d\beta_0}{d\tau}&=(r-q)\beta_1+\kappa_Q\gamma_Q\beta_2,\\ \frac{d\beta_1}{d\tau}&=0,\\ \frac{d\beta_2}{d\tau}&=-\frac{\beta_1}{2}-\kappa_Q\beta_2+\frac{1}{2}\left(\beta_1^2+2\rho\sigma\beta_1\beta_2+\sigma^2\beta_2^2\right).\end{aligned} \tag{24}$$

The characteristic function of the marginal density of the terminal value of $y=\log(S_T/K)$ requires the solution of Equations (24) with initial conditions $\beta_0(0)=0$, $\beta_1(0)=i\omega_Y$ and $\beta_2(0)=0$ for the particular values of ω_Y needed to construct the half-range and full-range Fourier series approximations of transitional probability density.

On a practical note, the fact that the characteristic function of $\log(S/K)$ embeds the difference between the risk-free rate of interest and the dividend rate suggest at first sight that the coefficients $\beta_0,\cdots,\beta_M$ must be computed whenever the risk-free rate or dividend rate changes, potentially each day, and not just when the parameters of the model are changed. The key to avoiding this difficulty is to note that the difference $(r-q)$ enters the calculation of β_0 alone, and that the behaviours of $\beta_1,\cdots,\beta_M$ are independent of the behaviour of β_0. Consequently, the resolution of this practical dilemma is to divide the calculation of β_0 into two stages with the first stage treating only those calculations that involve the difference $(r-q)$ and the second stage treating everything that is independent of the value of $(r-q)$. Both stages are brought together each day in the calculation of the expected payoff, but only the first stage of calculation needs be done on a day-to-day basis. In particular, this calculation is very easy as it amounts to adding $(r-q)T\omega_n$ to the second stage calculation of β_0 in the computation of $\chi(\omega_n)$.

6. ERROR ANALYSIS

Let $f_Q(y)$ denote the marginal density of $Y=\log(S/K)$, where K denotes the strike price of an asset. The purpose of this section is to demonstrate that the error in estimating the price of a call option using only that part of the density $f_Q(y)$ in the finite interval $[a,b]$ can be made arbitrarily small by choosing a suitably large interval. The function $f_Q(y)$, when truncated to the interval $[a,b]$, is assumed to satisfy the Dirichlet conditions and therefore is guaranteed to have a convergent Fourier series on $[a,b]$. The approximation procedure introduces three different errors which are now described.

6.1. Truncation Error

Truncation error occurs when the semi-infinite interval in Expression (13) or (16) is replaced by an integral over a finite interval, say $[0,b]$ in these expressions. With this approximation in place, the price of the call option is approximated by the expression

$$e^{-rT}K\int_0^b \left(e^y - 1\right) f_Q(y)\,dy \tag{25}$$

resulting in an error due to the loss of the contribution to the price from values of y in $[b,\infty]$. A straightforward analysis indicates that the error in replacing the true cost of the call option by Expression (25) is

$$e^{-rT}K\left[\int_0^\infty \left(e^y - 1\right) f_Q(y)\,dy - \int_0^b \left(e^y - 1\right) f_Q(y)\,dy\right]$$

which in turn simplifies to give

$$= e^{-rT}K\left[\left(e^b - 1\right)\left(1 - F_Q(b)\right) + \int_b^\infty e^y \left(1 - F_Q(y)\right) dy\right], \tag{26}$$

where $F_Q(y)$ is the cumulative function of y. Evidently the truncation error in pricing a call option is the sum of two positive contributions, both of which are driven by the extent to which the restriction of the marginal density of Y to the interval $[a,b]$ fails to capture density in the upper tail of the distribution of Y. Thus Expression (25) with a numerically accurate representation of the true marginal density of Y will always underestimate the true value of a call option, although the error can be made arbitrarily small by taking $[a,b]$ to be a suitably large interval.

The inference from this observation is that approximations of $f_Q(y)$ which place unit density in $[a,b]$ may potentially compensate for some of the error incurred in using the truncation procedure to price a call option. The half-range Fourier cosine and full-range Fourier series approximations of transitional probability density both fall into this category, which in turn suggests that the use of the truncation procedure with these approximations of marginal density may well provide rather smaller pricing errors than might casually be anticipated. By contrast, the half-range Fourier sine series approximation of marginal density does not capture unit density in the interval $[a,b]$, and the corresponding conjecture is that this approximation of marginal density will perform less well than the half-range Fourier cosine and full-range Fourier series approximations with regard to the pricing of a call option. Table 4 indicates that this conjecture is borne out in practice.

6.2. Approximation Errors

Suppose now that the transitional density $f_Q(y)$ in the interval $[a,b]$ is replaced by the Fourier series

$$f_Q(y)=\frac{a_0}{2}+\sum_{k=1}^{\infty}a_k\cos\omega_k(y-a)+\sum_{k=1}^{\infty}b_k\sin\omega_k(y-a), \tag{27}$$

where the choice of frequencies ω_k and the values of a_k, b_k depend on the choice of Fourier representation. Two errors arise in the calculation of Expression (25). First, the plan is to replace the coefficients a_k, b_k in Equation (27) by the misspecified coefficients A_k, B_k computed directly from the characteristic function $\chi(\omega_k)$ of $f_Q(y)$ via the formula

Table 4. Percentage errors recorded by the half-range cosine series, half-range sine series and full-range Fourier series of pricing a European call option and a binary option with strike K = 1200 and maturities one month (panel (a)), three months (panel (b)) and six months (panel (c)). The factor refers to the multiple of $\sigma\sqrt{T}$ used to establish the interval over which to compute the numerical approximation.

(a)

Factor	Call	Strike = 1200	T = 0.083	Binary	Strike = 1200	T = 0.083
	Cos	Sin	Full	Cos	Sin	Full
5.00	−4.160746	−17.315659	−7.719366	−0.031036	−4.731845	−1.295507
6.00	−0.000537	−0.129887	−0.052720	0.000369	−0.017647	−0.005300
7.00	0.008273	0.018872	0.001766	−0.002995	−0.000040	−0.000010
8.00	0.058873	0.019228	0.002259	0.000286	−0.000010	0.000003
9.00	0.015106	0.018981	0.002546	−0.000038	0.009813	0.000007

(b)

Factor	Call	Strike = 1200	T = 0.083	Binary	Strike = 1200	T = 0.083
	Cos	Sin	Full	Cos	Sin	Full
5.00	−0.006890	−0.084497	−0.035497	0.000000	−0.010412	−0.003842
6.00	−0.000014	−0.000471	−0.000204	0.000000	−0.000041	−0.000015
7.00	0.000000	0.000004	−0.000001	0.000000	−0.000000	0.000001
8.00	0.000000	0.000005	−0.000001	0.000000	−0.000000	0.000000
9.00	0.000000	0.000005	−0.000001	0.000000	0.000000	0.000000

(c)

Factor	Call	Strike = 1200	T = 0.500	Binary	Strike = 1200	T = 0.500
	Cos	Sin	Full	Cos	Sin	Full
5.00	−0.001412	−0.020338	−0.009446	0.000000	−0.002338	−0.000993
6.00	−0.000005	−0.000115	−0.000055	0.000000	−0.000010	−0.000004
7.00	0.000000	−0.000000	−0.000000	0.000000	0.000000	−0.000000
8.00	0.000000	−0.000000	−0.000000	0.000000	0.000000	−0.000000
9.00	0.000000	−0.000000	−0.000000	0.000000	0.000000	−0.000000

$$A_k + iB_k = \frac{2e^{-i\omega_k a}}{b-a}\chi(\omega_k). \tag{28}$$

Consequently a misspecification error arises in the Fourier coefficients. Second, the truncation of the Fourier series (27) to a finite number of terms, say N terms, introduces another approximation error. Therefore the price of a call option based on this strategy is

$$e^{-rT}K\left[\frac{A_0}{2}\int_0^b (e^y - 1)dy + \sum_{k=1}^{N} A_k \int_0^b (e^y - 1)\cos\omega_k (y-a)dy + \sum_{k=1}^{N} B_k \int_0^b (e^y - 1)\sin\omega_k (y-a)dy\right] \tag{29}$$

in which $f_Q(y)$ in Expression (25) has been replaced by the first $(N+1)$ terms of the Fourier series (27) with misspecified coefficients A_k and B_k. The error introduced by this approximation is

$$e^{-rT}K\int_0^b (e^y - 1) f_Q(y)dy - C \tag{30}$$

which has explicit expression

$$e^{-rT}K\left[\frac{(a_0 - A_0)}{2}\int_0^b (e^y - 1)dy + \sum_{k=1}^{N}(a_k - A_k)\int_0^b (e^y - 1)\cos\omega_k (y-a)dy + \sum_{k=1}^{N}(b_k - B_k)\int_0^b (e^y - 1)\sin\omega_k (y-a)dy + \sum_{k=N+1}^{\infty} a_k \int_0^b (e^y - 1)\cos\omega_k (y-a)dy + b_k \int_0^b (e^y - 1)\sin\omega_k (y-a)dy\right]. \tag{31}$$

The misspecification error in the coefficients a_k and b_k due to the use of the coefficients A_k and B_k respectively is determined from the identity

$$
\begin{aligned}
a_k + ib_k &= \frac{2}{b-a}\int_a^b f_Q(y)\mathrm{e}^{i\omega_k(y-a)}\mathrm{d}y \\
&= A_k + iB_k - 2\frac{\mathrm{e}^{-i\omega_k a}}{b-a}\left[\int_{-\infty}^a f_Q(y)\mathrm{e}^{i\omega_k y}\mathrm{d}y + \int_b^\infty f_Q(y)\mathrm{e}^{i\omega_k y}\mathrm{d}y\right].
\end{aligned}
$$

Standard properties of integral Calculus guarantee that

$$
\left|(a_k + ib_k) - (A_k + iB_k)\right| = \frac{2}{b-a}\left|\int_{-\infty}^a f_Q(y)\mathrm{e}^{i\omega_k y}\mathrm{d}y + \int_b^\infty f_Q(y)\mathrm{e}^{i\omega_k y}\mathrm{d}y\right| \le \varepsilon_1,
$$

Where

$$
\varepsilon_1 = \frac{2}{b-a}\left[\int_{-\infty}^a f_Q(y)\mathrm{d}y + \int_b^\infty f_Q(y)\mathrm{d}y\right] = \frac{2}{b-a}\left[F_Q(a) + 1 - F_Q(b)\right]. \tag{32}
$$

It therefore follows directly that

$$
|a_k - A_k| \le \varepsilon_1, \quad |b_k - B_k| \le \varepsilon_1, \tag{33}
$$

for all values of k. Thus the misspecification error in the Fourier coefficients is driven entirely by the choice of interval $[a,b]$ via the magnitude of the cumulative function $F_Q(y)$ at $y=a$ and $y=b$. In practice this error will be significantly smaller than the maximum bound given in Equation (33) as a result of arithmetical cancellation due to the oscillatory nature of the integrand, but as commented previously, the decay of this error as $[a,b] \to \mathbb{R}$ will be determined by the behaviour of $F_Q(y)$ as $y \to -\infty$ and as $y \to \infty$.

In conclusion, the total error in pricing a call option, namely

$$\left| e^{-rT} K \int_0^{\infty} \left(e^y - 1 \right) f_Q(y) dy - C \right|, \tag{34}$$

is formed by connecting together Equation (26) for the error arising in the truncation of the density $f_Q(y)$ to the finite interval $[a, b]$, Equation (29) for the value of the option in terms of the misspecified Fourier coefficients A_k, B_k, and Equation (31) for the size of the resulting misspecification error in the option price. The result is that the error is

$$e^{-rT} K \left[\left(e^b - 1 \right) \left(1 - F_Q(b) \right) + \int_b^{\infty} e^y \left(1 - F_Q(y) \right) dy \right] + e^{-rT} K \left[\frac{(a_0 - A_0)}{2} \int_0^b \left(e^y - 1 \right) dy \right.$$
$$+ \sum_{k=1}^{N} (a_k - A_k) \int_0^b \left(e^y - 1 \right) \cos \omega_k (y - a) dy + \sum_{k=1}^{N} (b_k - B_k) \int_0^b \left(e^y - 1 \right) \sin \omega_k (y - a) dy$$
$$\left. + \sum_{k=N+1}^{\infty} a_k \int_0^b \left(e^y - 1 \right) \cos \omega_k (y - a) dy + \sum_{k=N+1}^{\infty} b_k \int_0^b \left(e^y - 1 \right) \sin \omega_k (y - a) dy \right]. \tag{35}$$

Each integral is replaced by its value and the triangle inequality is used to deduce that

$$\begin{aligned}|\text{Error}| \le \mathrm{e}^{-rT} K \Bigg[& \left(\mathrm{e}^{b}-1\right)\left(1-F_{Q}(b)\right)+\int_{b}^{\infty} \mathrm{e}^{y}\left(1-F_{Q}(y)\right) \mathrm{d} y+\frac{\varepsilon_{1}}{2}\left(\mathrm{e}^{b}-b-1\right) \\ & +\varepsilon_{1} \sum_{k=1}^{N}\left|\frac{\left(\sin \omega_{k} a+\omega_{k} \cos \omega_{k} a\right)}{\omega_{k}\left(1+\omega_{k}^{2}\right)}-\frac{\mathrm{e}^{b} \cos \omega_{k}(b-a)}{1+\omega_{k}^{2}}\right| \\ & +\varepsilon_{1} \sum_{k=1}^{N}\left|\frac{\mathrm{e}^{b} \omega_{k} \cos \omega_{k}(b-a)-\sin \omega_{k} a}{1+\omega_{k}^{2}}+\frac{\cos \omega_{k} a-\cos \omega_{k}(b-a)}{\omega_{k}\left(1+\omega_{k}^{2}\right)}\right| \\ & +\sum_{k=N+1}^{\infty}\left|a_{k}\right|\left|\frac{\sin \omega_{k} a+\omega_{k} \cos \omega_{k} a}{\omega_{k}\left(1+\omega_{k}^{2}\right)}-\frac{\mathrm{e}^{b} \cos \omega_{k}(b-a)}{1+\omega_{k}^{2}}\right| \\ & +\sum_{k=N+1}^{\infty}\left|b_{k}\right|\left|\frac{\mathrm{e}^{b} \omega_{k} \cos \omega_{k}(b-a)-\sin \omega_{k} a}{1+\omega_{k}^{2}}+\frac{\cos \omega_{k} a-\cos \omega_{k}(b-a)}{\omega_{k}\left(1+\omega_{k}^{2}\right)}\right|\Bigg] . \end{aligned} \tag{36}$$

The contributions to the error from the first, second, third and fourth terms on the right hand side of inequality (36) are dominated by the behaviour of $F_Q(a)$ and $1-F_Q(b)$. By choosing the interval $[a,b]$ suitably large, $F_Q(a)$ and $1-F_Q(b)$ can be made arbitrarily small. The behaviour of the fifth and sixth terms on the right hand side of inequality (36) depends on the choice of N. Bearing in mind that $\omega_k = O(k)$, then it straightforward to show that there are positive constants C_1 and C_2 such that

$$\left|\frac{\left(\sin \omega_{k} a+\omega_{k} \cos \omega_{k} a\right)}{\omega_{k}\left(1+\omega_{k}^{2}\right)}-\frac{\mathrm{e}^{b} \cos \omega_{k}(b-a)}{1+\omega_{k}^{2}}\right| \le \frac{\mathrm{e}^{b}+1}{\omega_{k} \sqrt{1+\omega_{k}^{2}}} \le \frac{C_{1}}{k^{2}},$$

$$\left|\frac{\mathrm{e}^{b} \omega_{k} \cos \omega_{k}(b-a)-\sin \omega_{k} a}{1+\omega_{k}^{2}}+\frac{\cos \omega_{k} a-\cos \omega_{k}(b-a)}{\omega_{k}\left(1+\omega_{k}^{2}\right)}\right| \le \frac{\mathrm{e}^{b} \sqrt{1+\omega_{k}^{2}}+1}{\omega_{k} \sqrt{1+\omega_{k}^{2}}} \le \frac{C_{2}}{k}.$$

The well known result that if $f_Q(y)$ is a continuous function3 of y, then $|a_k| < K/k$ and $|b_k| < K/k$ indicates that the fourth and fifth terms of inequality (36) are $O\left(k^{-3}\right)$ and $O\left(k^{-2}\right)$ respectively. Thus the fifth and sixth terms on the right hand side of inequality (36) can also be made arbitrarily small by a suitably large choice of N .

In conclusion, the error in pricing a call option can be made arbitrarily small by restricting the marginal probability density to a finite interval $[a,b]$ and approximating the density in that interval by a misspecified Fourier series constructed from the characteristic function at the appropriate frequencies.

7. PERFORMANCE UNDER SIMULATION

A series of simulation experiments was undertaken in order to examine the efficacy of the half-range Fourier cosine series, half-range Fourier sine series and full-range Fourier series in respect of how accurately these approximations price European call options and binary options. The first experiment prices options in a Black- Scholes world so that a closed-form solution may be used to assess the pricing error, whereas the second experiment prices the same options using Heston's model of stochastic volatility.

7.1. Black-Scholes Pricing

Assume that the asset price, S, follows a geometric Brownian motion

$$\mathrm{d}S = S\left(r\mathrm{d}t + \sigma \mathrm{d}W\right)$$

in which $\mathrm{d}W$ is the increment of a Wiener process. The advantage of using this specification is that, following the seminal work of Black and Scholes [13] , exact prices are known for each type of option for all combinations of spot price S_0, strike price K , and maturity T . It follows, therefore, that the relative percentage pricing error incurred using numerical methods based on Fourier series for a variety of different strike prices and maturities can be identified exactly.

Three major experiments are performed. In each of these experiments $r = 0.05$, $\sigma = 0.15$ per year, $S_0 = 1000$ and the strikes K are taken to be 1200, 1000 and 800 respectively so as to examine the performance of the algorithms when the options are out-of-the-money, at-the-money and in-the-money. In each simulation the accuracy of the various approximations of transitional density is assessed for prescribed values of a and b in which the size of the interval $[a,b]$ is expressed as a multiple of $\sigma\sqrt{T}$ ranging from 5 to 9. The results of these exercises are reported in Tables 4-6 with each table reporting the results for maturities of one month (panel (a)), three months (panel (b)) and six months (panel (c)).

Two very clear general conclusions emerge from these results.

1) For options that are deep in-the-money (Table 6) in which $S_0 = 1000$ and $K = 800$, or at-the-money (Table 5) in which $S_0 = 1000$ and K = 1000, it matters little (if at all) which of these three pricing algorithms are chosen irrespective of the maturity of the option. Indeed the relative pricing error is zero for all practical purposes.

2) For options that are deep out-of-the-money (Table 4) in which $S_0 = 1000$ and $K = 1200$, the choice of algorithm is more important. The results may be summarized succinctly as follows.

a) The half-range Fourier sine series does not perform as well as the other two approximations and its use is therefore not recommended. This finding accords well with our previous intuition based on a consideration of the contribution to the price of a call option lost as a result of truncating marginal density.

b) The half-range Fourier cosine series and the full-range Fourier series both perform relatively well. When the size of the interval of approximation is a relatively small multiple of $\sigma\sqrt{T}$, namely either 5 or 6, then the half-range Fourier cosine series performs better that the full-range Fourier series. As the multiple of $\sigma\sqrt{T}$ increases and the size of the interval of approximation becomes larger, the full-range Fourier series begins to dominate. When the interval of approximation has size $10\sigma\sqrt{T}$, then the full-range Fourier series is unambiguously superior to the half-range Fourier cosine series, particularly for options of short maturity.

On the basis of this analysis and on accuracy grounds, it is hard to ignore the claim that the full-range Fourier series is the algorithm of choice when using Fourier methods to price options. Moreover, the full-range Fourier series converges faster than either the half-range sine or cosine series and is therefore likely to price options more rapidly. These themes are explored

Table 5. Percentage errors recorded by the half-range cosine series, half-range sine series and full-range Fourier series of pricing a European call option and a binary option with strike K = 1000 and maturities one month (panel (a)), three months (panel (b)) and six months (panel (c)). The factor refers to the multiple of $\sigma\sqrt{T}$ used to establish the interval over which to compute the numerical approximation.

(a)

Factor	Call			Binary		
	Strike = 1000		T = 0.083	Strike = 1000		T = 0.083
	Cos	Sin	Full	Cos	Sin	Full
5.00	−0.000044	−0.000313	−0.001045	0.000000	−0.000159	−0.000043
6.00	−0.000000	−0.000002	−0.000005	0.000000	−0.000001	−0.000000
7.00	0.000000	0.000000	0.000000	0.000000	−0.000000	−0.000000
8.00	−0.000000	0.000000	0.000000	0.000000	−0.000000	−0.000000
9.00	−0.000000	0.000000	0.000000	0.000000	0.000000	0.000000

(b)

Factor	Call			Binary		
	Strike = 1000		T = 0.250	Strike = 1000		T = 0.250
	Cos	Sin	Full	Cos	Sin	Full
5.00	−0.000063	−0.001375	−0.000547	0.000000	−0.000202	−0.000074
6.00	−0.000000	−0.000007	−0.000003	0.000000	−0.000001	−0.000000
7.00	0.000000	−0.000000	−0.000000	0.000000	−0.000000	−0.000000
8.00	−0.000000	−0.000000	−0.000000	0.000000	−0.000000	−0.000000
9.00	−0.000000	−0.000000	−0.000000	0.000000	0.000000	0.000000

(c)

Factor	Call Strike = 1000 T = 0.500			Binary Strike = 1000 T = 0.500		
	Cos	Sin	Full	Cos	Sin	Full
5.00	−0.000116	−0.002208	−0.001059	−0.000000	−0.000254	−0.000108
6.00	−0.000000	−0.000012	−0.000006	−0.000000	−0.000001	−0.000000
7.00	0.000000	−0.000000	−0.000000	−0.000000	−0.000000	−0.000000
8.00	0.000000	−0.000000	0.000000	−0.000000	−0.000000	−0.000000
9.00	0.000000	−0.000000	0.000000	−0.000000	−0.000000	−0.000000

in more detail in the pricing of call options under Heston's model of stochastic volatility.

7.2. Heston's Model

A total of approximately 40,000 options over ten years were generated by simulation of Heston's model. Sixteen options were generated each day spread over 4 maturities ranging from 92 to 5 days and 4 strikes two of which are initialised at 20% out of and into the money and two of which are initialised at 7% out of and into the money. The half-range Fourier cosine series and the full-range Fourier series are then used to price call options and binary options with these strikes. In this instance no exact solutions are available, and so the accuracy of each method in respect of each type of option is gauged by comparison against values calculated using a large interval. The left hand and middle columns of Table 7 show respectively the L_2 and L_1 relative pricing errors for out-of-the-money call options calculated using the half-range Fourier cosine series and the full-range Fourier series. The right hand column of Table 7 shows the CPU time (secs) needed to perform the calculations. The left hand column specifies the length of the interval $[a,b]$ in multiples of $\sigma_Y = \sqrt{\gamma_Q DT}$, the unconditional standard deviation of the distribution of Y about $\log(S_0/K)$ after an interval of duration DT, taken to be a trading day in the simulation study, i.e. $DT = 1/252$.

A clear finding from Table 7 is that the half-range Fourier cosine series performs well in respect of the L_2 measure for shorter intervals and in the L_1 measure for almost all choices of interval length. By contrast the performance of the full-range Fourier series is poor with regard to both measures for shorter intervals. However, the quality of approximation provided by the half-range cosine series is erratic as the size of the Fourier window increases whereas that provided by the full-range Fourier series improves systematically to the extent that its performance surpasses that of the half-range Fourier cosine series for intervals of length 24 standard deviations or more. Furthermore, this level of accuracy is achieved by the full-range Fourier series in approximately 25% quicker than that required by the half-range Fourier cosine series.

Table 6. Percentage errors recorded by the half-range cosine series, half-range sine series and full-range Fourier series of pricing a European call option and a binary option with strike K = 800 and maturities one month (panel (a)), three months (panel (b)) and six months (panel (c)). The factor refers to the multiple of $\sigma\sqrt{T}$ used to establish the interval over which the numerical approximation is computed.

(a)

Factor	Call	Strike = 800	T = 0.083	Binary	Strike = 800	T = 0.083
	Cos	Sin	Full	Cos	Sin	Full
5.00	−0.000003	−0.000184	−0.000049	−0.000043	−0.000127	−0.000068
6.00	−0.000000	−0.000001	−0.000000	−0.000000	−0.000000	−0.000000
7.00	−0.000000	0.000000	−0.000000	−0.000000	−0.000000	0.000000
8.00	−0.000000	0.000000	−0.000000	−0.000000	−0.000000	0.000000
9.00	−0.000000	0.000000	0.000000	−0.000000	−0.000010	0.000000

(b)

Factor	Call	Strike = 800	T = 0.250	Binary	Strike = 800	T = 0.250
	Cos	Sin	Full	Cos	Sin	Full
5.00	−0.000011	−0.000343	−0.000133	−0.000000	−0.000111	−0.000041
6.00	−0.000000	−0.000002	−0.000001	−0.000000	−0.000000	−0.000000
7.00	−0.000000	0.000000	−0.000000	−0.000000	−0.000000	0.000000
8.00	−0.000000	0.000000	−0.000000	−0.000000	−0.000000	0.000000
9.00	−0.000000	0.000000	0.000000	−0.000000	−0.000000	0.000000

(c)

Factor	Call	Strike = 800	T = 0.500	Binary	Strike = 800	T = 0.500
	Cos	Sin	Full	Cos	Sin	Full
5.00	−0.000022	−0.000580	−0.000259	−0.000008	−0.000147	−0.000063
6.00	−0.000000	−0.000003	−0.000001	−0.000000	−0.000001	−0.000000
7.00	−0.000000	0.000000	−0.000000	−0.000000	−0.000000	0.000000
8.00	−0.000000	0.000000	−0.000000	−0.000000	−0.000000	0.000000
9.00	−0.000000	0.000000	0.000000	−0.000000	−0.000000	0.000000

Table 7. The L_2 and L_1 relative pricing errors calculated from the half-range Fourier cosine series and the full-range Fourier series for out-of-the-money call options based on Heston's model. The timings refer to the times (seconds) needed to price 12 daily call options over 10 years with maturities ranging from 5 to 92 days. The factor refers to the multiple of σ_Y used to establish the interval over which the numerical approximation is computed.

Factor	Out-of-the-money call (L_2 Error)		Out-of-the-money call (L_1 Error)		Timings (sec)	
	Cos Series	Full Series	Cos Series	Full Series	Cosine	Full
10	0.000228	0.172011	0.000289	0.208449	0.140	0.109
12	0.000523	0.056627	0.000045	0.063078	0.156	0.109
14	0.000053	0.016726	0.000029	0.018403	0.187	0.141
16	0.000242	0.004654	0.000030	0.005235	0.203	0.172
18	0.000374	0.001251	0.000030	0.001468	0.234	0.171
20	0.000190	0.000330	0.000035	0.000417	0.265	0.187
22	0.000064	0.000086	0.000028	0.000129	0.281	0.218
24	0.000460	0.000023	0.000030	0.000052	0.296	0.218
26	0.000058	0.000010	0.000027	0.000031	0.328	0.249
28	0.000315	0.000008	0.000029	0.000025	0.343	0.265
30	0.000219	0.000007	0.000035	0.000021	0.359	0.280

Table 8. The L_2 and L_1 relative pricing errors calculated from the half-range cosine series and the full-range Fourier series for in-the-money call options based on Heston's model. The timings refer to the times (seconds) needed to price 12 daily call options over 10 years with maturities ranging from 5 to 92 days. The factor refers to the multiple of σ_Y used to establish the interval over which the numerical approximation is computed.

Factor	In-the-money call (L_2 Error)		In-the-money call (L_1 Error)		Timings (sec)	
	Cos Series	Full Series	Cos Series	Full Series	Cosine	Full
10	0.000027	0.002421	0.000362	0.215297	0.140	0.109
12	0.000008	0.000845	0.000063	0.061658	0.156	0.109
14	0.000008	0.000281	0.000043	0.017301	0.187	0.141
16	0.000008	0.000090	0.000043	0.004790	0.203	0.172
18	0.000008	0.000029	0.000043	0.001326	0.234	0.171
20	0.000024	0.000011	0.000052	0.000382	0.265	0.187
22	0.000008	0.000008	0.000041	0.000130	0.281	0.218
24	0.000008	0.000008	0.000043	0.000062	0.296	0.218
26	0.000007	0.000007	0.000041	0.000044	0.328	0.249
28	0.000008	0.000007	0.000042	0.000038	0.343	0.265
30	0.000025	0.000007	0.000053	0.000034	0.359	0.280

Fang and Oosterlee [10] suggest choosing intervals of length 20 standard deviations. In this problem, it is evident that the half-range Fourier cosine series still enjoys an advantage over the full-range Fourier series for intervals of this length. The suggestion of this investigation is that 20 standard deviations should be regarded as a minimum length of interval. Table 8 shows the L_2 and L_1 relative pricing errors calculated from the half- range Fourier cosine series and the full-range Fourier series in respect of in-the-money call options based on Heston's model.

The results reported in Table 8 exhibit a similar pattern of behaviour to those reported in Table 7, namely that for intervals of short length the half-range Fourier cosine series outperforms the full-range Fourier series with respect to both the L_2 and L_1 measures of accuracy, but that this dominance vanishes when the length of the interval of approximation is 24 standard deviations or more. In particular, both approaches generate noticeably smaller relative errors for in-the-money options than for out-of-the-money options. Because in-the-money call options carry a higher price, one inference of this observed reduction in relative error is that the absolute pricing error in using each Fourier representation is insensitive to the moneyness of the option. Table 8 and Table 9report the performance of the half-range Fourier cosine series and the full-range Fourier series when used to price binary options.

The results presented in Table 9 and Table 10 demonstrate that both the half-range Fourier cosine series and the full-range Fourier series both generate good estimates of the price of binary options with the former performing better than the latter for intervals of short length, but with this advantage disappearing when the length of the interval is increased.

The results reported in Tables 7-10 are characterised by two common denominators. First, the computational time required by the half-range Fourier cosine series is uniformly longer than that required by the full-range Fourier series for an interval of fixed length. The simple explanation for this observation is that the full-range Fourier series uses larger frequencies for a given length of interval, and therefore the full-range Fourier series converges more rapidly. Second, the pricing of call options and binary options using the half-range Fourier cosine series representation of transitional density is noticeably better than the corresponding pricing using the full-range Fourier series for short intervals $[a,b]$, but this advantage vanishes in the case of the binary option when the interval becomes suitably large and is reversed in the case

Table 9. The L_2 and L_1 relative pricing errors calculated from the half-range cosine series and the full-range Fourier series for out-of-the-money binary options based on Heston's model. The timings refer to the times (seconds) needed to price 12 daily call options over 10 years with maturities ranging from 5 to 92 days. The factor refers to the multiple of σ_Y used to establish the interval over which the numerical approximation is computed.

Factor	Binary out-of-the-money (L_2 Error)		Binary out-of-the-money (L_1 Error)		Timings (sec)	
	Cos Series	Full Series	Cos Series	Full Series	Cosine	Full
10	0.000020	0.005846	0.000000	0.000328	0.172	0.094
12	0.000006	0.001753	0.000000	0.000081	0.203	0.125
14	0.000020	0.000504	0.000000	0.000020	0.234	0.125
16	0.000007	0.000141	0.000000	0.000005	0.265	0.141
18	0.000028	0.000047	0.000000	0.000001	0.281	0.156
20	0.000004	0.000022	0.000000	0.000000	0.312	0.172
22	0.000020	0.000020	0.000000	0.000000	0.343	0.187
26	0.000020	0.000020	0.000000	0.000000	0.390	0.203
30	0.000022	0.000020	0.000000	0.000000	0.437	0.250

Table 10. The L_2 and L_1 relative pricing errors calculated from the half-range cosine series and the full-range Fourier series for in-the-money binary options based on Heston's model. The timings refer to the times (seconds) needed to price 12 daily call options over 10 years with maturities ranging from 5 to 92 days. The factor refers to the multiple of σ_Y used to establish the interval over which the numerical approximation is computed.

Factor	Binary in-the-money		Binary in-the-money		Timings (sec)	
	(L_2 Error)	(L_2 Error)	(L_1 Error)	(L_1 Error)		
	Cos Series	Full Series	Cos Series	Full Series	Cosine	Full
10	0.000025	0.000678	0.000004	0.000186	0.172	0.109
12	0.000007	0.000193	0.000000	0.000044	0.203	0.125
14	0.000007	0.000054	0.000000	0.000011	0.234	0.125
16	0.000007	0.000016	0.000000	0.000003	0.265	0.156
18	0.000007	0.000008	0.000000	0.000001	0.281	0.156
20	0.000006	0.000007	0.000000	0.000000	0.312	0.171
22	0.000009	0.000007	0.000000	0.000000	0.343	0.187
26	0.000016	0.000007	0.000000	0.000000	0.390	0.219
30	0.000012	0.000007	0.000000	0.000000	0.437	0.234

of a call option. The primary explanation for this behaviour lies in the realisation that tail density is characterised by long wavelengths, or equivalently by the presence of low frequency terms, whereas peaky density is characterised by short wavelengths, or equivalently high frequency terms.

For an interval of given length, the frequencies present in the half-range Fourier cosine expansion are smaller that the frequencies present in the full-range Fourier series, and therefore when the length of the interval $[a,b]$ is small, the half-range Fourier cosine series better captures tail behaviour. Increasing the length of the interval lowers the range of frequencies admitted to the full-range Fourier series expansion, which in turn improves its ability to handle tail density. However, the same increase in length will benefit the half-range Fourier cosine series representation of transitional density only provided there remains significant tail density to be characterised by these new frequencies. On the other hand, the half-range Fourier cosine series will always suffer from the drawback that the (periodic) function it represents has gradient zero at $x=a$ and $x=b$ (because the series is not generally differentiable at these points). Consequently, the half-range Fourier cosine series approximation will always misrepresent the gradient of the true density function at $x=a$ and $x=b$ in contrast to the behaviour of the approximation provided by the full-range Fourier series.

8. CONCLUSION

A. S. Hurn, K. A. Lindsay, A. J. McClelland One clear conclusion from these calculations is that the half-range Fourier cosine series and the full-range Fourier series perform uniformly better than the half-range Fourier sine series. The half-range Fourier cosine series and the full-range Fourier series

both perform with credit. When the length of the interval is relatively small, say ten or so standard deviations, it is clear that the half-range Fourier cosine series outperforms the full-range Fourier series over the same interval. On the other hand for intervals of larger length the full-range Fourier series is at least as good as the half-range Fourier cosine series, and outperforms the latter in pricing out-of-the-money call options, in particular, with maturities of three months or less. In general the full-range

Fourier series outperforms the half-range Fourier cosine series in all circumstances provided that the interval is suitably large, although the effect is so small as not to be significant in practice. The explanation for this behaviour lies in the fact that the half-range Fourier cosine series, although not representing a differentiable function, always uses a lower spectrum of frequencies than the full-range Fourier series and therefore enjoys an initial advantage in describing tail density. As the interval length is increased, this advantage vanishes. On the other hand the larger spectrum of the full-range Fourier series guarantees more rapid convergence. Of course, timing issues may not be important if small numbers of model call prices are to be calculated, but when facing calibration problems or estimation problems as described in the introductory remarks, timing issues are indeed significant once adequate numerical accuracy is assured.

REFERENCES

1. Johannes, M.S., Polson, N.G. and Stroud, J.R. (2009) Optimal Filtering of Jump Diffusions: Extracting Latent States from Asset Prices. Review of Financial Studies, 22, 2759-2799.
2. Broadie, M., Chernov, M. and Johannes, M. (2007) Model Specification and Risk Premia: Evidence from Futures Options. Journal of Finance, 62, 1453-1490.
3. Christoffersen, P., Jacobs, K. and Mimouni, K. (2010) Volatility Dynamics for the S&P500: Evidence from Realized Volatility, Daily Returns and Option Prices. Review of Financial Studies, 23, 3141-3189.
4. Hurn, A.S., Lindsay, K.A. and McClelland, A.J. (2012) Estimating the Parameters of Stochastic Volatility Models Using Option Price Data. Unpublished Working Paper, NCER.
5. Andersen, T.G., Fusari, N. and Todorov, V. (2012) Parametric Inference and Dynamic State Recovery from Option Pa- nels. NBER Working Paper Series.
6. Carr, P.P. and Madan, D.B. (1999) Option Evaluation Using the Fast Fourier Transform. Journal of Computational Fi- nance, 2, 61-73.
7. Borak, S., Detlefsen, K. and Hardle, W. (2005) FFT Based Option Pricing. SFB Discussion Paper 649.
8. Lord, R., Fang, F. Bervoets, F. and Oosterlee, C.W. (2007) A Fast and Accurate FFT-Based Methodology for Pricing Early-Exercise

Options under Levy Processes. SIAM Journal of Scientific Computing, 20, 1678-1705.

9. Kwok, Y.K., Leung, K.S. and Wong, H.Y. (2012) Efficient Options Pricing Using the Fast Fourier Transform. In: Duan, J.C., Ed., Handbook of Computational Finance, Springer, Berlin, 579-604.

10. Fang, F. and Oosterlee, C.W. (2008) A Novel Pricing Method for European Options Based on Fourier-Cosine Series Expansions. SIAM Journal on Scientific Computing, 31, 826-848.

11. Zhang, B., Grzelak, L.A. and Oosterlee, C.W. (2012) Efficient Pricing of Commodity Options with Earlyexercise under the Ornstein-Uhlenbeck Process. Applied Numerical Mathematics, 62, 91-111.

12. Heston, S.L. (1993) A Closed-Form Solution for Options with Stochastic Volatility with Applications to Bond and Cu- rrency Options. Review of Financial Studies, 6, 327-343.

13. Black, F. and Scholes, M. (1973) The Pricing of Options and Corporate Liabilities. Journal of Political Economy, 81, 637-654.

NOTES

[1]The Finite Fourier Transform is the integral expression defining the coefficients of a Fourier series.

[2]Historically, half-range Fourier series have largely arisen as analytical tools for handling various types of boundary conditions when solving partial differential equations using integral transforms.

[3]Sharper results can be obtained if $f_Q(y) \in C^p[a,b]$ for $p > 1$. However the function represented by the half-range Fourier cosine series is not generally differentiable at $x = a$, and so the convergence argument is based on the weakest condition satisfied by Fourier coefficients.

Index